Synthesis Lectures on Ocean Systems Engineering

Series Editor

Nikolas Xiros, University of New Orleans, New Orleans, LA, USA

The series publishes short books on state-of-the-art research and applications in related and interdependent areas of design, construction, maintenance and operation of marine vessels and structures as well as ocean and oceanic engineering.

Alexander Arnfinn Olsen

Subsea Pipeline Systems

Building and Classification

 Springer

Alexander Arnfinn Olsen (iD)
Southampton, UK

ISSN 2692-4420   ISSN 2692-4471 (electronic)
Synthesis Lectures on Ocean Systems Engineering
ISBN 978-3-031-74789-2   ISBN 978-3-031-74790-8 (eBook)
https://doi.org/10.1007/978-3-031-74790-8

© The Editor(s) (if applicable) and The Author(s), under exclusive license to Springer
Nature Switzerland AG 2025, corrected publication 2025

This work is subject to copyright. All rights are solely and exclusively licensed by the Publisher, whether the whole or part of the material is concerned, specifically the rights of translation, reprinting, reuse of illustrations, recitation, broadcasting, reproduction on microfilms or in any other physical way, and transmission or information storage and retrieval, electronic adaptation, computer software, or by similar or dissimilar methodology now known or hereafter developed.
The use of general descriptive names, registered names, trademarks, service marks, etc. in this publication does not imply, even in the absence of a specific statement, that such names are exempt from the relevant protective laws and regulations and therefore free for general use.
The publisher, the authors and the editors are safe to assume that the advice and information in this book are believed to be true and accurate at the date of publication. Neither the publisher nor the authors or the editors give a warranty, expressed or implied, with respect to the material contained herein or for any errors or omissions that may have been made. The publisher remains neutral with regard to jurisdictional claims in published maps and institutional affiliations.

This Springer imprint is published by the registered company Springer Nature Switzerland AG
The registered company address is: Gewerbestrasse 11, 6330 Cham, Switzerland

If disposing of this product, please recycle the paper.

Preface

This guide applies to the classification of design, construction, and installation of off-shore pipelines made of metallic materials, as well as the periodic surveys required for the maintenance of classification. Serviceability of pipelines is also addressed, but only to the extent that proper functioning of the pipe and its components affects safety. This guide may also be used for certification or verification of design, construction, and installation of pipelines. Class will typically certify or verify the design, construction, and installation of offshore pipelines when requested by the asset owner or as mandated by government regulations to verify compliance with the prevailing regulations and requirements as covered by specific technical requirements, national standards, or other applicable industry standards.

This guidance has been developed for worldwide application, and as such, the satisfaction of individual requirements may require comprehensive data, analyses and plans to demonstrate adequacy. This especially applies for pipelines located in frontier areas, such as those characterised by relatively great water depth or areas with little or no previous operating experience. Conversely, many provisions of this guide often can be satisfied merely on a comparative basis of local conditions or past successful practices. The author acknowledges that a wide latitude exists as to the extent and type of documentation which is required for submission to satisfy this guide. It is not the intention of this guide to impose requirements or practices in addition to those that have previously proven satisfactory in similar situations.

Where available, design requirements in this guide have been posed in terms of existing methodologies and their attendant safety factors, load factors, or permissible stresses that are deemed to provide an adequate level of safety. Primarily, the class's use of such methods and limits in this guide reflects what is considered to be the current state of practice in offshore pipeline design. At the same time, it is acknowledged that methods of design, construction, and installation are constantly evolving. In recognition of these facts, the guide specifically allows for such innovations and the appendices are intended to reflect this. The application of this guide will not seek to inhibit the use of any technological approach that can be shown to produce an acceptable level of safety.

Southampton, UK   Alexander Arnfinn Olsen

The original version of the book has been revised. A correction to this book can be found at https://doi.org/10.1007/978-3-031-74790-8_17

Contents

List of Tables

Scope and Conditions of Classification

1

1.1   General

The intention of this guide is to serve as technical documentation for design, fabrication, installation and maintenance of offshore production, transfer and export pipelines made of metallic materials. The principal objectives are to specify the minimum requirements for classing, continuance of classing, certification and verification by Class. In addition to the requirements outlined in this guide, the design of a marine system requires consideration of all relevant factors related to its functional requirements and long-term integrity, such as:

- Compliance with local laws, acts and regulations,
- Functional requirements,
- Physical site information; and
- Operational requirements.

Pipelines which have been built, installed, tested and commissioned to the satisfaction of Class Surveyors to the full requirements of this guide or to its equivalent, where approved by another recognised authority, may be classed and distinguished in the asset's record by

✠ A1 Offshore installation – offshore pipelines

or equivalent.

Pipelines which have not been built, installed, tested and commissioned under Class survey, but which are submitted to Class for classification, will typically be subjected to a special classification survey. Where found satisfactory, and thereafter approved by Class,

© The Author(s), under exclusive license to Springer Nature Switzerland AG 2025   1
A. A. Olsen, *Subsea Pipeline Systems*, Synthesis Lectures on Ocean Systems
Engineering, https://doi.org/10.1007/978-3-031-74790-8_1

the asset may be classed and distinguished in the record in the manner as described above, however any especial distinguishing marks (for example, the mark ✠,signifying survey during construction) may be omitted in accordance with the Class Rules.

Data on the pipeline will be published in the asset's Record as to the latitude and longitude of its location, type, dimensions and depth of water at the site.

The requirements outlined in this guide are applicable to those features that are permanent in nature and can be verified by plan review, calculation, physical survey or other appropriate means. Any statement in this guide regarding other features should be considered as guidance only to the asset designer, builder, owner, et al.

1.2   Risk Evaluations for Alternative Arrangements and Novel Features

Risk assessment techniques may be used to demonstrate that alternatives and novel features provide acceptable levels of safety in line with current offshore and marine industry practice. It is strongly recommended readers refer to the appropriate Class Rules pertaining to risk assessment applications for the marine and offshore industries provides guidance on how to prepare a risk evaluation to demonstrate equivalency or acceptability for a proposed design.

Risk evaluations for the justification of alternative arrangements or novel features may be applicable either to the installation as a whole, or to individual systems, subsystems or components. Class will usually consider the application of risk evaluations for alternative arrangements and novel features for subsea pipeline systems.

Portions of the subsea pipeline system or any of its components thereof not explicitly included in the risk evaluation submitted to Class are to comply with any applicable part of the Class Rules and guides. If any proposed alternative arrangement or novel feature affects any applicable requirements of Flag and Coastal State, it is the responsibility of the asset owner to discuss with the applicable authorities the acceptance of alternatives based on risk evaluations.

For new or novel concepts, i.e. applications or processes that have no previous experience in the environment being proposed, the guidance encompassed in the Class Rules may not be directly applicable to them. The process described in this guide draws upon engineering, testing and risk assessments in order to determine if the concept provides acceptable levels of safety in line with current industry practices.

For the classing of pipelines, the documentation submitted to Class must usually include reports, calculations, drawings and other documentation necessary to demonstrate the adequacy of the design of the pipelines. More often than not, the required documentation is to include those items listed in this chapter.

1.3   Plans and Specifications

Plans and specifications depicting or describing the arrangements and details of the major items of pipelines are to be submitted for review or approval in a timely manner. These include:

- Site plan indicating bathymetric features along the proposed route, the location of obstructions to be removed, the location of permanent man-made structures, the portions of the pipe to be buried and other important features related to the characteristics of the sea floor,
- Structural plans and specifications for pipelines, their supports and coating,
- Schedules of non-destructive testing and quality control procedures,
- Flow diagram indicating temperature and pressure profiles; and
- Specifications and plans for instrumentation and control systems and safety devices.

When requested by the asset owner, the asset owner and Class may jointly establish a schedule for information submission and plan approval. This schedule, to which Class will usually aim to adhere to as far as reasonably possible, is to reflect the fabrication and construction schedule and the complexity of the pipeline systems as they affect the time required for review of the submitted data.

1.4   Information Memorandum

Dependent on specific Class requirements, an information memorandum on pipelines may need to be prepared and submitted to Class. In these situations, Class will review the contents of the memorandum to establish consistency with other data submitted for the purpose of obtaining classification or certification. Where required, the information memorandum should contain, as appropriate to the pipelines, any/all of the following:

- A site plan indicating the general features at the site and the field location of the pipelines,
- Environmental design criteria, including the recurrence interval used to assess environmental phenomena,
- Plans showing the general arrangement of the pipelines,
- Description of the safety and protective systems provided,
- Listing of governmental authorities having authority over the pipelines,
- Brief description of any monitoring proposed for use on the pipelines; and
- Description of manufacturing, transportation and installation procedures.

1.5   Site-Specific Conditions

An environmental condition report is to be submitted, describing anticipated environmental conditions during pipe laying, as well as environmental conditions associated with normal operating conditions and the design environmental condition. Items to be assessed are to include, as appropriate, waves, current, temperature, tide, marine growth, chemical components of air and water, ice conditions, earthquakes and other pertinent phenomena.

A route investigation report is to be submitted, addressing with respect to the proposed route of the pipeline system the topics of seafloor topography and geotechnical properties. In the bathymetric survey, the width of the survey along the proposed pipeline route is to be based on consideration of the expected variation in the final route in comparison with its planned position, and the accuracy of positioning devices used on the vessels employed in the survey and in the pipe laying operation. The survey is to identify, in addition to bottom slopes, the presence of any rocks or other obstructions that might require removal, gullies, ledges, unstable slopes and permanent obstructions, such as existing man-made structures. The geotechnical properties of the soil are to be established to determine the adequacy of its bearing capacity and stability along the route. The methods of determining the necessary properties are to include a suitable combination of in-situ testing, seismic survey, and boring and sampling techniques. As appropriate, soil testing procedures are to adequately assess sea floor instability, scour or erosion and the possibility that soil properties may be altered due to the presence of the pipe, including reductions in soil strength induced by cyclic soil loading or liquefaction. The feasibility of performing various operations relative to the burial and covering of the pipe is to be assessed with respect to the established soil properties.

Where appropriate, data established for a previous installation in the vicinity of the pipeline proposed for classification may be utilised, if acceptable to Class.

1.6   Material Specifications

Documentation for all materials of the major components of pipelines is to indicate that the materials satisfy the requirements of the pertinent specification. For line pipes, specifications are to identify the standard with which the product is in complete compliance, the size and weight designations, material grade and class, process of manufacture, heat number and joint number. Where applicable, procedures for storage and transportation of the line pipes from the fabrication and coating yards to the offshore destination are to be given. Material tests, if required, are to be performed to the satisfaction of Class.

1.7   Design Data and Calculations

Information is to be submitted for the pipelines that describe the material data, models and variability, long-term degradation data and models, methods of material system selection, analysis and design that were employed in establishing the design. The estimated design life of the pipelines is to be stated. Where model testing is used as the basis for a design, the applicability of the test results will depend on the demonstration of the adequacy of the methods employed, including enumeration of possible sources of error, limits of applicability and methods of extrapolation to full scale data. It is preferable that the procedures be reviewed and agreed upon before material and component model testing is performed. Calculations are to be submitted to demonstrate the adequacy of the proposed design and are to be presented in a logical and well-referenced fashion, employing a consistent system of units. Where suitable, at least the following calculations are to be performed.

1.8   Structural Strength and On-Bottom Stability Analysis

Calculations are to be performed to demonstrate that, with respect to the established loads and other influences, the pipelines, support structures and surrounding soil possess sufficient strength and on-bottom stability with regard to failure due to the following:

- Excessive stresses and deflections,
- Fracture,
- Fatigue,
- Buckling,
- Collapse; and
- Foundation movements.

Additional calculations may be required to demonstrate the adequacy of the proposed design. Such calculations are to include those performed for unusual conditions and arrangements, as well as for the corrosion protection system.

1.9   Installation Analysis

With regard to the installation procedures, installation analyses, including trenching effects, are to be submitted for review. These calculations demonstrate that the anticipated loading from the selected installation procedures does not jeopardise the strength and integrity of the pipelines.

1.10   Safety Devices

An analysis of the pipeline safety system may be submitted to demonstrate compliance with API RP 14G, or equivalent. As a recommended minimum, the following safety devices are to be part of the pipelines:

- For departing pipelines, a high-low pressure sensor is required on the floater or platform to shut down the wells, and a check valve is required to avoid backflow,
- For incoming pipelines an automatic shutdown valve is to be connected to the floater or platform's emergency shutdown system, and a check valve is required to avoid backflow; and
- For bi-directional pipelines, a high-low pressure sensor is required on the floater or platform to shut down the wells, and an automatic shutdown valve is to be connected to the floater or platform's emergency shutdown system.

Shortly after the pipelines are installed, all safety systems are to be checked in order to verify that each device has been properly installed and calibrated and is operational and performing as prescribed. In the post-installation phase, the safety devices are to be tested at specified regular intervals and periodically operated so that they do not become fixed by remaining in the same position for extended periods of time.

1.11   Installation Manual

Where required, a manual is to be submitted describing procedures to be employed during the installation of pipelines and is as a minimum to include:

- List of the tolerable limits of the environmental conditions under which pipe laying may proceed,
- Procedures and methods to evaluate impact and installation damage tolerance,
- Procedures to be followed should abandonment and retrieval be necessary,
- Repair procedures to be followed should any component of pipelines be damaged during installation; and
- Contingency plan.

An installation manual is to be prepared to demonstrate that the methods and equipment used by the contractor meet the specified requirements. As a minimum, the qualification of the installation manual is to include procedures related to:

- Quality assurance plan and procedures,
- Welding procedures and standards,

- Welder qualification,
- Non-destructive testing procedures,
- Repair procedures for field joints, internal and external coating repair, as well as repair of weld defects, including precautions to be taken during repairs to prevent overstressing the repair joints,
- Qualification of pipe-lay facilities, such as tensioner and winch,
- Start and finish procedure,
- Laying and tensioning procedures,
- Abandonment and retrieval procedures,
- Subsea tie-in procedures,
- Intervention procedures for crossing design, specification and construction, bagging, permanent and temporary support design, specification and construction, etc.,
- Trenching procedures,
- Burying procedures,
- Field joint coating and testing procedures,
- Drying procedures; and
- System pressure test procedures and acceptance criteria.

Full details of the lay vessel, including all cranes, abandonment and recovery winches, stinger capacities and angles, welding and non-destructive testing gear, firing line layout and capacity and vessel motion data are to be provided, together with general arrangement drawings showing plans, elevations and diagrams of the pipeline assembly, welding, non-destructive testing, joint coating and lay operations. Full details of any trenching and burying equipment are to be provided.

1.12   Pressure Test Report

A report including procedures for and records of the testing of each pipeline system is to be submitted. The test records are, as a minimum, to include an accurate description of the facility being tested, the pressure gauge readings, the recording gauge charts, the dead weight pressure data and the reasons for and disposition of any failures during a test. A profile of the pipeline that shows the elevation and test sites over the entire length of the test section is to be included. Records of pressure tests are also to contain the names of the asset owner and the test contractor, the date, time and test duration, the test medium and its temperature, the weather conditions and sea water and air temperatures during the test period. Plans for the disposal of test medium together with discharge permits may be required to be submitted to Class.

1.13   Operations Manual

Where required, an operations manual is to be prepared to provide a detailed description of the operating procedures to be followed for expected conditions. The operations manual is to include procedures to be followed during start-up, operations, shutdown conditions and anticipated emergency conditions. This manual is to be submitted to Class for record and file.

1.14   Maintenance Manual

Where required, a maintenance manual providing detailed procedures for how to ensure the continued operating suitability of the pipeline system is to be submitted to Class for approval. The manual is, as a minimum, to include provisions for the performance of the following items:

• Visual inspection of non-buried parts of pipelines to verify that no damage has occurred to the systems and that the systems are not being corroded,
• Evaluation of the cathodic protection system performance by potential measurements,
• Detection of dents and buckles by caliper pigging; and
• Inspection and testing of safety and control devices.

Additionally, Class may require gauging of pipe thickness should it be ascertained that pipelines are undergoing erosion or corrosion. Complete records of inspections, maintenance and repairs of pipelines are to be provided for Class.

1.15   As-Built Documents

The results of surveys and inspections of the pipelines are to be given in a report which, as a minimum, is to include the following details:

• Plot of the final pipeline position, superimposed on the proposed route including pipeline spans and crossings,
• Description and location of any major damage to the pipelines alongside information regarding how such damage was repaired; and
• Description of the effectiveness of burial operations (if applicable for pipelines).

As appropriate, results of additional inspections, which may include those for the proper operation of corrosion control systems, fibre optic and/or damage sensors, buckle detection by caliper pig or other suitable means and the testing of alarms, instrumentation and safety and emergency shutdown systems, may be included.

Survey, Inspection, and Testing Regimes

2.1 General

This chapter relates to the inspection and survey of pipelines at different phases including fabrication, installation, and testing after installation. The phases of fabrication and construction covered by this chapter include pipe and coating manufacture, fabrication, assembly, and line pipe pressure test. The phases of installation include route survey of the pipelines, preparation, transportation, field installation, construction, system pressure test and survey of the as-built installation. The post-installation phase includes survey for continuance of classification, accounting for damage, failure, and repair.

2.1.1 Quality Control and Assurance Programmes

A quality control and assurance programme compatible with the type, size and intended functions of pipelines is to be developed and submitted to Class for review. Class will review, approve and, as necessary, request modification of this programme. The contractor is to work with Class to establish the required hold points on the quality control programme to form the basis for all future inspections at the fabrication yard and surveys of the pipeline. As a minimum, the items enumerated in the various applicable sections below are to be covered by the quality control programme. If required, Surveyors may be assigned to monitor the fabrication of pipelines and assure that competent personnel are carrying out all tests and inspections specified in the quality control programme. It is to be noted that the monitoring provided by Class is a supplement to and not a replacement for inspections to be carried out by the asset constructor or operator. The quality control programme, as appropriate, is to include the following items:

© The Author(s), under exclusive license to Springer Nature Switzerland AG 2025
A. A. Olsen, *Subsea Pipeline Systems*, Synthesis Lectures on Ocean Systems
Engineering, https://doi.org/10.1007/978-3-031-74790-8_2

- Material quality and test requirements,
- Line pipe manufacturing procedure specification and qualification,
- Welder qualification and records,
- Pre-welding inspection,
- Welding procedure specifications and qualifications,
- Weld inspection,
- Tolerances and alignments,
- Corrosion control systems,
- Concrete weight coating,
- Non-destructive testing,
- Inspection and survey during pipe laying,
- Final inspection and system pressure testing,
- Pigging operations and tests, and
- Final as-built condition survey and acceptance.

2.1.2 Access and Notification

During fabrication and construction Class representatives may need access to pipelines at all reasonable times. In these situations, Class will need to be notified as to when and where line pipe, pipeline and pipeline components may be examined. Where Class finds occasion to recommend repairs or further inspection, notice will usually be made to the contractor or their representatives.

2.1.3 Identification of Materials

The asset contractor is to maintain a data system of material for line pipe, pipeline components, joints, anodes, and coatings. Data concerning place of origin and results of relevant material tests are to be retained and made readily available during all stages of construction.

2.2 Inspection and Testing in Fabrication Phase

Specifications for quality control programmes of inspection during fabrication of line pipe and pipeline components are given in this section. Qualification tests are to be conducted to document that the requirements of the specifications are satisfied.

2.2.1   Material Quality

The physical properties of the line pipe material and welding are to be consistent with the specific application and operational requirements of pipelines. Suitable allowances are to be added for possible degradation of the physical properties in the subsequent installation and operation activities. Verification of the material quality is to be done by the Surveyor at the manufacturing plant, in accordance with Chaps. 3 and 4. Alternatively, materials manufactured to the recognised standards or proprietary specifications may be accepted by Class, provided such standards give acceptable equivalence with the requirements of this guide.

2.2.2   Manufacturing Procedure Specification and Qualification

A manufacturing specification and qualification procedure is to be submitted for acceptance before production start. The manufacturing procedure specification is to state the type and extent of testing, the applicable acceptance criteria for verifying the properties of the materials and the extent and type of documentation, record, and certificate. All main manufacturing steps from control of received raw material to shipment of finished line pipe, including all examination and checkpoints, are to be described. Class will survey formed line pipe, pipeline, pipeline components such as bends, tees, valves, etc., for their compliance with the dimensional tolerances, chemical composition and mechanical properties required by the design.

2.2.3   Welder Qualification and Records

Welders who are to work on pipelines must be qualified in accordance with the welder qualification tests specified in a recognised code, such as API STD 1104 and Section IX of the ASME "Boiler and Pressure Vessel Code". Certificates of qualification are to be prepared to record evidence of the qualification of each welder qualified by an approved standard/code. In the event welders have been previously qualified, in accordance with the requirements of a recognised code, and provided that the period of effectiveness of previous testing has not lapsed, these welder qualification tests may be accepted.

2.2.4   Pre-welding Inspection

Prior to welding, each pipe is to be inspected for dimensional tolerance, physical damage, coating integrity, interior cleanliness, metallurgical flaws and proper fit-up and edge preparation.

2.2.5   Welding Procedure Specifications and Qualifications

Welding procedures are to conform to the provisions of a recognised code, such as API STD 1104, or the asset owner's specifications. A written description of all procedures previously qualified may be employed in the construction, provided it is included in the quality control programme and made available to Class. When it is necessary to qualify a welding procedure, this is to be accomplished by employing the methods specified in the recognised code. All welding is to be based on welding consumables and welding techniques proven to be suitable for the types of material, pipe, and fabrication in question. As a minimum, the welding procedure specification is to contain the following items:

- Base metal and thickness range,
- Types of electrodes,
- Joint design,
- Weld consumable and welding process,
- Welding parameters and technique,
- Welding position,
- Preheating, and
- Inter pass temperatures and post weld heat treatment.

For underwater welding, additional information is to be specified, if applicable, including water depth, pressure and temperature, product composition inside the chamber and the welding umbilical and equipment.

2.2.6   Weld Inspection

As part of the quality control programme, a detailed plan for the inspection and testing of welds is to be prepared. The physical conditions under which welding is to proceed, such as weather conditions, protection and the condition of welding surfaces are to be noted. Alterations in the physical conditions may be required should it be established that satisfactory welding cannot be obtained. Where weld defects exceed the acceptability criteria, they are to be completely removed and repaired. Defect acceptance criteria may be project-specific, as dictated by welding process, non-destructive testing resolution and results of fatigue crack growth analysis. The repaired weld is to be re-examined using acceptable non-destructive methods.

2.2.7 Tolerances and Alignments

The dimensional tolerance criteria are to be specified in developing the line pipe man-ufacturing specification. Inspections and examinations are to be carried out to ensure that the dimensional tolerance criteria are being met. Particular attention is to be paid to the out-of-roundness of pipes for which buckling is an anticipated failure mode. Struc-tural alignment and fit-up prior to welding are to be monitored to ensure the consistent production of quality welds.

2.2.8 Corrosion Control Systems

The details of any corrosion control system employed for pipelines are to be submitted for review. Installation and testing of the corrosion control systems are to be carried out in accordance with the approved plans and procedures. Where employed, the application and resultant quality of corrosion control coatings (external and internal) are to be inspected to ensure that specified methods of application are followed and that the finished coating meets specified values for thickness, lack of holidays (small parts of the structural sur-faces unintentionally left without coating), hardness, etc. Visual inspection, micrometre measurement, electric holiday detection or other suitable means are to be employed in the inspection.

2.2.9 Concrete Weight Coatings

Weight coatings applied when onshore or, if applicable, when on the lay vessel are to be inspected for compliance with the specified requirements for bonding, strength and hard-ness, weight control and any necessary special design features. Production tests should be carried out at regular intervals to prove compliance with the specifications.

2.2.9.1 Non-destructive Testing (NDT)

A system of non-destructive testing is to be included in the fabrication and construc-tion specification of pipelines. The minimum extent of non-destructive testing is to be in accordance with this guide or a recognised design Code. All non-destructive testing records are to be reviewed and approved by Class. Additional non-destructive testing may be requested if the quality of fabrication or construction is not in accordance with industry standards.

2.2.9.2 Fabrication Records

A data book of the record of fabrication activities is to be developed and maintained so to compile as complete a record as is practicable. The pertinent records are to be adequately prepared and indexed to assure their usefulness, and they are to be stored in a manner that is easily recoverable. As a minimum, the fabrication record is to include, as applicable, the following:

- Manufacturing specification and qualification procedures records,
- Material trace records (including mill certificates),
- Welding procedure specification and qualification records,
- Welder qualification,
- Non-destructive testing procedures and operator's certificates,
- Weld and non-destructive testing maps,
- Shop welding practices,
- Welding inspection records,
- Fabrication specifications,
- Structural dimension check records,
- Non-destructive testing records,
- Records of completion of items identified in the quality control programme,
- Assembly records,
- Pressure testing records,
- Coating material records,
- Batch No., etc.,
- Concrete weight coating mix details, cube test, etc., and
- The compilation of these records is a condition of certifying pipelines.

After fabrication and assembly, these records are to be retained by the asset operator or fabricator for future reference. The minimum time for record retention is not to be less than the greatest of the following:

- Warranty period,
- Time specified in fabrication and construction agreements, and/or
- Time required by statute or governmental regulations.

2.3   Inspection and Testing During Installation

This section gives the specifications and requirements for the installation phase, covering route survey of pipelines prior to installation, installation manual, installation procedures, contingency procedures, as-laid survey, system pressure test, final testing, and preparation for operation.

2.3.1 Specifications and Drawings for Installation

The specifications and drawings for installation are to be detailed and prepared giving the descriptions of and requirements for the installation procedures to be employed. The requirements are to be available in the design premise, covering the final design, verification and acceptance criteria for installation and system pressure test, records, and integrity of pipelines. The drawings are to be detailed enough to demonstrate the installation procedures step-by-step. The final installation results are to be included in the drawings.

2.3.2 Installation Manual

Qualification of the installation manual is specified in Chap. 1, under the heading "Installation manual."

2.3.3 Inspection and Survey During Pipe Laying

Representatives from Class may be required to witness the installation of pipelines to ensure that it proceeds according to approved procedures.

2.3.4 Final Inspection and Pressure Testing

A final inspection of the installed pipeline is to be completed to verify that it satisfies the approved specifications used in its fabrication and the requirements of this guide. If the pipeline is to be buried, inspection will normally be required both before and after burial operations. As appropriate, additional inspections, which may include those for the proper operation of corrosion control systems, buckle detection by calliper pig or other suitable means, the testing of alarms, instrumentation, safety systems and emergency shutdown systems, are to be performed.

2.3.5 Inspection for Special Cases

Areas of pipelines may require inspection after one of the following occurrences:

- Environmental events of major significance,
- Significant contact from surface or underwater craft, dropped objects or floating debris,
- Any evidence of unexpected movement, and

- Any other conditions which might adversely affect the stability, structural integrity, or safety of pipelines.

Damage that affects or may affect the integrity of pipelines is to be reported at the first opportunity by the asset operator for examination by Class. All repairs deemed necessary by Class are to be carried out to their satisfaction.

2.3.6   Notification

The asset operator is to notify Class on all occasions when parts of pipelines not ordinarily accessible are to be examined. If during any visit the Surveyor should find occasion to recommend repairs or further examination, this is to be made known to the asset operator immediately in order that appropriate action may be taken.

2.4   In-Service Inspection and Survey

The phases of operation include operation preparation, inspection, survey, maintenance, and repair. During the operation condition, in-service inspections and surveys are to be conducted for pipelines. In-service inspections and surveys are to be planned to identify the actual conditions of pipelines for the purpose of integrity assessment. In-service inspection can be planned based on the following:

- At each Annual Survey, the records of maintenance are to be reviewed for compliance with the approved maintenance plan. The function of the safety protective devices is to be proven in order,
- Any subsea maintenance inspection carried out internally or externally of the pipeline is to be verified and reported by the Class attending Surveyor, and
- At each five (5) year interval, the complete maintenance records are to be reviewed and any major inspections, in accordance with the approved maintenance plans, are to be witnessed and reported by the Class attending Surveyor.

2.5   Inspection for Extension of Use

Existing pipelines to be used at the same location for an extended period beyond the original design life are to be subject to additional structural inspection to identify the actual condition of the pipelines. The extent of the inspection will depend on the completeness of the existing inspection documents. Any alterations, repairs, replacements, or installation of equipment since installation are to be included in the records. The inspection schedule of the pipelines can be planned based on the requalification or reassessment of the systems applying, e.g., structural reliability methodology and incorporating past inspection records.

Materials and Welding

3

3.1　General

This chapter specifies the line pipe material requirements, including steel pipes and other special metallic pipes used for pipeline applications. Material and dimensional standards for metallic pipe are to be in accordance with this guide with respect to chemical composition, material manufacture, tolerances, strength and testing requirements. A specification is to be prepared stating the requirements for materials and for manufacture, fabrication and testing of line pipes, including their mechanical properties.

3.2　Metallic Pipe

The line pipe materials used under this guide are to be carbon steels, alloy steels or other special materials, such as titanium, manufactured according to a recognised standard. The materials are to be able to maintain the structural integrity of pipelines for hydrocarbon transportation under the effects of service temperature and anticipated loading conditions. Materials in near vicinity are to be qualified in accordance with applicable specifications for chemical compatibility. The following aspects are to be considered in the selection of material grades:

- Mechanical properties,
- Internal fluid properties and service temperature,
- Resistance to corrosion effects,
- Environmental and loading conditions,
- Installation methods and procedure,
- Weight requirement,

A. A. Olsen, *Subsea Pipeline Systems*, Synthesis Lectures on Ocean Systems Engineering, https://doi.org/10.1007/978-3-031-74790-8_3

- Weldability,
- Fatigue and fracture resistance.

Documentation for items such as formability, welding procedure, hardness, toughness, fatigue, fracture, and corrosion characteristics is to be submitted for Class review to substantiate the applicability of the proposed materials.

3.3   Steel Line Pipe

The material, dimensional standards and manufacturing process of steel pipe are to be in accordance with API SPEC 5L, ISO 3183–1~3 or other recognised standards. Approval by Class is required for the intended application with respect to chemical composition, material manufacture, tolerances, strength and testing requirements.

3.3.1   Chemical Composition

The chemical composition of line pipes, as determined by heat analysis, is to conform to the applicable requirements of the grade and type of steel material. However, the requirements of chemical composition may be agreed upon between the asset operator and the line pipe manufacturer.

3.3.2   Weldability

The carbon equivalent (C_{eq}) and the cold cracking susceptibility (P_{cm}) for evaluating the weldability of steel pipes may be calculated from the ladle analysis, in accordance with the following equations (percentage of weight):

$$C_{eq} = C + \frac{Mn}{6} + \frac{Cr + Mo + V}{5} + \frac{Ni + Cu}{15}$$

$$P_{cm} = C + \frac{Si}{30} + \frac{Mn + Cu + Cr}{20} + \frac{Ni}{60} + \frac{Mo}{15} + \frac{V}{10} + 5B$$

Selection of C_{eq} and P_{cm}, , as well as their maximum values, is to be agreed between the asset operator and the steel mill when the steel is ordered to ensure weldability. When low carbon content is used for sour service, the value of the cold cracking susceptibility (P_{cm}) is to be limited. However, the behaviour of steel pipe during and after welding is dependent on the steel, the filler metals used and the conditions of the welding process. Unless it can be documented otherwise, a testing programme is to be performed to qualify candidate line pipe materials and filler metals.

3.3.3   Pipe Manufacturing Procedure

During the initial stages of manufacture of each item (after this called "first day production"), certain supplementary tests and qualification of manufacturing and testing facilities will be required in addition to the testing and inspection required during production of pipe. This testing and qualification are also to be carried out if there are any alterations in the manufacturing, testing or inspection procedures that might result in a detrimental change in pipe quality. No pipe will be accepted until first day production tools and qualifications are accepted. The fabrication procedures are to comply with an approved standard pertinent to the type of pipe being manufactured. All non-destructive testing operations referred to in this chapter are to be conducted by non-destructive testing personnel qualified and certified in accordance with standards such as ASNT SNT-TC-1A, ISO 9712 or other applicable codes.

The manufacturer is to prepare a manufacturing procedure specification for review by Class. The manufacturing procedure specification is to document the forming techniques and procedures, welding procedures and welding testing, material identification, mill pressure testing, dimensional tolerances, surface conditions and properties to be achieved and verified. Pipes are to be selected from initial production for manufacturing procedure qualification through mechanical, corrosion and non-destructive testing. Deepwater service requires the manufacture, inspection, testing and shipping of line pipes with minimum requirements as follows:

- The steel is to be fully killed and fine grain,
- Plate is to be manufactured to a well-known and documented practice. All heat-treating facilities are to be equipped with instrumentation such that all temperatures can be controlled and recorded, and
- All production welding is to be automatic.

Pipe may be either non-expanded or cold expanded. Cold expanded pipe is not to exceed 2.0 percent maximum expansion, nor is it to exceed the amount of expansion used during first day production tests by more than 0.2 percent. Pipe may be cold compressed. Cold compressed pipe is not to exceed 2.0 percent maximum compression, nor is it to exceed the amount of compression used during first day production tests. The plates and or pipe from each heat are to remain segregated during the entire manufacturing, testing, inspection, and shipping process, as is practical.

3.3.4   Fabrication Tolerance

The fabrication tolerance may be agreed upon between the operator and the line pipe manufacturer but is to be consistent with the design requirements. The pipes may be sized to their final dimensions by expansion and straightening. The pipes are to be delivered to the dimensions specified in the manufacturing procedure.

3.3.5 Fracture Arrest Toughness

Fracture toughness values for crack arrest given in ISO codes and API RP are adequate for design factors up to 72% and are given mainly for land-based pipelines. The acceptance criteria for fracture arrest toughness of offshore pipelines are to be agreed upon between the operator and line pipe manufacturer.

3.3.6 Mill Pressure Test

The mill test pressure and duration may be agreed upon between the asset operator and the line pipe manufacturer, but it is to be consistent with the design requirements. The mill pressure test is to be conducted after final pipe expansion and straightening.

3.4 Line Pipe Materials for Special Applications

This section defines the minimum requirements for line pipe materials such as carbon steel, stainless steel, duplex, clad carbon steel and titanium alloy for extreme temperatures, sour service, or other special applications.

3.4.1 Sour Service

Line pipe materials for sour (H_2S-containing) service are to satisfy the criteria of NACE MR0175 for resistance to sulphide stress cracking (SSC) and hydrogen-induced cracking (HIC) failures. Materials that are not listed in NACE MR0175 are to be tested according to procedures NACE TM0177 and NACE TM0284 for both materials and welds. The acceptance criteria are to be agreed upon between the asset Operator and the line pipe manufacturer based on the intended service condition.

3.4.2 Stainless, Duplex, and Super Duplex Stainless-Steel Pipes

The chemical composition and the manufacturing of stainless-steel pipes are to follow standards such as ASTM A790. The manufacturer is to establish the manufacturing pro-cedure for the pipes, which is to contain relevant information about steel manufacturing, pipe manufacturing, welding and control methods which are to follow recognised standards such as API SPEC 5LC. Mechanical tests are to be performed after heat treatment, expansion, and final shaping. Specific tests may be required to meet specific project requirements.

3.4.3 Clad Pipe

Clad pipes are to be compatible with the functional requirements and service conditions as specified for the project. Material dimensional standards and manufacturing process of clad steel pipe are to be in accordance with API SPEC 5LD or equivalent recognised standards.

3.4.4 Titanium Pipe

Specific compositional limits and tensile property minimums for titanium alloy tubular products may be produced in accordance with ASTM B861 and ASTM B862 specifications. Titanium alloys are highly corrosion-resistant to produced well fluid, including all hydrocarbons, acidic gases (CO_2 and H_2S), elemental sulphur and sweet and sour chloride brines at elevated temperatures. Titanium alloys are also generally resistant to well, drilling and completion fluids.

3.5 Marking, Documentation and Transportation

Pipes are to be properly marked for identification by the manufacturer. The marks are to identify the standard with which the product is in complete compliance, the size and weight designations, material grade and class, process of manufacture, heat number and joint number. Pipe storage arrangements are to preclude possible damage, such as indentations of the surface and edges of pipes. Materials are to be adequately protected from deleterious influences during storage. The temperature and humidity conditions for storing weld filler material and coating are to comply with those specified in their controlling material specification or manufacturer's supplied information. Documentation for all materials of the major components of pipelines is to indicate that the materials satisfy the requirements of the pertinent specification. Material tests are to be performed to the satisfaction of Class. Procedure for the transportation of the line pipes from the fabrication and coating yards to the offshore destination is to be established. Transportation of the pipes is to follow the guidelines of API RP 5L1 and API RP 5LW or equivalent.

Pipe Components and Pipe Coating

4

4.1 General

The design of the pipeline includes various piping components. Specifications for each piping component and coating material used on a pipeline system are to be identified. The specifications are to be submitted to Class for approval if the components have special service conditions or deviate from the standards indicated in this guide or other comparable codes. For valves, fittings, connectors, and joints, if the wall thickness and yield strength between the adjoining ends are different, the joint design for welding is to be made in accordance with ASME B31.4, Fig. 434.8.6(a)-(2), for liquid pipelines, or ASME B31.8, Appendix I, Figure I5, for gas pipelines. The internal diameter of pipeline components is to be equal to that of the connecting pipeline sections. Consideration is to be given to effects of erosion at locations where the flow changes direction. Seal design for valves, fittings and connectors are to account for external hydrostatic pressure.

4.2 Piping Components

The piping components are to be suitable for the pipeline design conditions and be compatible with the line pipes in material, corrosion, and welding.

4.2.1 Flanges

Pipe flanges used for offshore pipelines vary depending on the connection requirement subsea and at the surface to the platforms. Typical flange materials and dimensions are to follow ASME B16.5, API SPEC 17D, and MSS SP-44, where applicable. The flange

© The Author(s), under exclusive license to Springer Nature Switzerland AG 2025
A. A. Olsen, *Subsea Pipeline Systems*, Synthesis Lectures on Ocean Systems
Engineering, https://doi.org/10.1007/978-3-031-74790-8_4

design may be determined by calculations in accordance with Section VIII of the ASME Boiler and Pressure Vessel Code.

4.2.2 Pipe Fittings

Pipe fittings are to match the design of the line pipes and flanges. Typical materials and dimensions are to follow ASME B16.9, B16.11, B16.25, MSS SP-75, and API SPEC 17D, where applicable.

4.2.3 Gaskets

Gaskets are to match the design of the flanges. Typical materials and dimensions are to follow ASME B16.20 and API SPEC 6A, where applicable.

4.2.4 Bolting

The bolting is to match the design of the flanges. Typical materials, dimensions and bolting torque are to follow ASME B16.5 and API SPEC 6A, where applicable.

4.2.5 Valves

The valves are to match the line pipes and flanges. Typical materials and dimensions are to follow ASME B16.34, API STD 600, and API SPEC 6D.

4.2.6 Subsea Tees

Subsea tees are to be of the extruded outlet, integral reinforcement type. The design is to be in accordance with ASME B 31.4, ASME B31.8 or equivalent codes or standards.

4.2.7 Y-pieces

Y-pieces and tees where the axis of the outlet is not perpendicular to the axis of the run are to be designed by finite element analysis.

4.2.8   Bends

Mitred bends are not permitted in offshore liquid and gas pipeline systems. Pipe that has been cold- worked solely for the purpose of increasing the yield strength to meet specified minimum yield strength is not allowed in offshore pipeline systems. Bends are to be made in such a manner as to preserve the cross-sectional shape of the pipe, and are to be free from buckling, cracks, or other evidence of mechanical damage. The pipe diameter is not to be reduced at any point by more than 2.5% of the nominal diameter, and the completed bend is to be able to pass the specified sizing pig.

4.2.9   Piping Supports and Foundation

Piping support and foundation design is to follow the appropriate criteria of ASME B31.4 and API RP 2A-WSD. Supporting elements, such as supports, braces and anchors for pipelines are to be designed in accordance with ASME B31.4 for liquid pipelines and ASME B31.8 for gas pipelines. No supporting elements are allowed to be welded directly to the pipeline except clamps, which are to fully encircle the pipe and be welded to the pipe by a full encirclement weld.

4.3   Pipe Coating

Specifications for corrosion protection coatings and concrete weight coatings are to be submitted to Class for approval if special service conditions exist. The weight coating specification is as a minimum to include:

- Chemical composition,
- Physical and strength properties, and
- Quality control procedures and verifying tests for manufacturing or production.

4.3.1   Corrosion Protection Coating

Corrosion protection coating materials are to be suitable for the intended use and consideration is to be given to:

- Corrosion protective properties,
- Temperature resistance,
- Adhesion and disbonding properties in conjunction with cathodic protection,
- Mechanical properties,

- Impact resistance,
- Durability,
- Shear strength,
- Tensile strength,
- Sea water resistance,
- Water absorption,
- Dielectric resistance,
- Compatibility with cathodic protection system,
- Resistance to chemical, biological and microbiological effects,
- Aging, brittleness, and cracking,
- Variation of properties with temperature and time, and
- Health and safety information and instruction according to national regulations.

The coating procedure is to comply with appropriate standards and is to include the details of the pipe surface preparation, production parameters, material specifications, application and testing methods, including acceptance criteria, and details of cutback lengths and coating termination. Before and after the coating application, inspection and testing are to be conducted by means of holiday detection to identify discontinuities or other defects that may impair its performance.

4.3.2 Weight Coating

Weight coating is, when applicable, to be applied to ensure vertical and horizontal on-bottom stability by providing negative buoyancy to the pipeline. The weight coating specification is to include:

- Mechanical properties, including strength, density, durability, etc.,
- Cement materials or equivalent,
- Reinforcement, including type, amount, and grade,
- Concrete coating method to achieve homogeneous and adequately consolidated coating,
- Curing method compatible with coating application,
- Repairs of uncured or hardened defective concrete coatings, and
- Storage, handling and transportation of coated pipe.

Inspection and testing are to be carried out at regular intervals during weight coating application, and consideration is to be given to:

- Mix proportions and water-cement ratio,
- Concrete density and compressive strength,

- Weight before and after concrete application,
- Outer diameter of coated pipe,
- Water absorption, and
- Compatibility with corrosion protection coating.

Stress concentration in the pipeline due to the weight coating is to be examined to avoid local damage in the form of buckling or fracture during handling and laying operations.

4.3.3 Insulation Coating

Thermal insulation coatings may be required for pipelines, spools, pipe-in-pipe systems, and pipeline bundle systems to ensure flow assurance, in which case, a design and qualification programme is to be submitted to Class for review. The thermal insulation design is to consider the coating material properties, including:

- Thermal conductivity,
- Density,
- Adhesion to base material,
- Abrasion resistance,
- Service pressure and temperature,
- Impact resistance,
- Creep,
- Durability against chemical, physical or biological effects,
- Water absorption, and
- Degradation during service.

Inspection is to be conducted both during surface preparation and after coating application.

4.3.4 Field Joint Coating

Field joint coating is to be placed on the pipe joint after completion of the welding and weld testing. Installation, inspection and testing procedures for the field joint are to be developed and submitted to Class.

4.4   Welding of Pipes and Piping Components

The welding of metallic pipes is to be performed in accordance with approved welding procedures that have been qualified to produce sound, ductile welds of adequate strength and toughness. Welding standards comparable to API STD 1104 and Section IX of the ASME Boiler and Pressure Vessel Code are to be employed in association with this guide. For special pipe materials, the applicability of the API STD 1104 is to be examined and verified at all stages of welding, and any alternative methods are to be submitted for review. Welders are to be tested in accordance with the welder qualification tests specified in recognised national codes, such as API STD 1104. Certificates of qualification are to be prepared to cover each welder when they are qualified by standards other than those of Class, and such certificates are to be available for the reference of the Surveyors. Before construction begins, details of the welding procedures and sequences are to be submitted for review. The details are to include:

- Base metal and thickness range,
- Types of electrodes,
- Edge preparation,
- Electrical characteristics,
- Welding technique,
- Proposed position and speed, and
- Preheating and post-weld heat treatment practices.

Welding procedures conforming to the provisions of an acceptable code may be qualified in the presence of the Surveyor, in accordance with the pertinent code. A written description of all prequalified procedures employed in the pipeline's construction is to be prepared and made available to the Surveyors. When it is necessary to qualify a welding procedure, this is to be accomplished by employing the methods specified in an acceptable code, and in the presence of the Surveyor.

4.5   Corrosion Control

A corrosion control system analysis is to be performed to determine necessary protection measures and to provide in-service performance criteria and procedure for maintaining the system. The analysis is to be submitted to Class for review and approval. This chapter recommends guidelines for the establishment of corrosion mitigation procedures for offshore pipelines. The following publications are incorporated by reference for the detection and mitigation of external and internal corrosion:

- ASME B31.4, Chapter VIII, and/or
- ASME B31.8, Chapter VI.

4.5.1 External Corrosion Control

Adequate anti-corrosion coating and cathodic protection are to be provided for protection against external corrosion and may include a galvanic anode system, an impressed current system or both. Design considerations are to be given to:

- Pipe surface area,
- Environmental condition,
- Suitability of galvanic anode systems under given marine environment,
- Design life of galvanic anode systems,
- How to minimise potential damage to the cathodic protection system during the lifecycle,
- Interference of electrical currents from nearby structures,
- Necessity of insulating joints for electrical isolation of portions of the system, and
- Inspection requirements for rectifiers or other impressed current sources.

4.5.2 Internal Corrosion Control

Adequate measures are to be taken against internal corrosion. Proper selection of pipe material, internal coating, injection of a corrosion inhibitor or a combination of such options are to be considered. When necessary, internal corrosion may be mitigated by the following:

- Running scrapers,
- Dehydration,
- Injection of corrosion inhibitors,
- Use of bactericides,
- Use of oxygen scavengers,
- Use of internal coating compatible to the contents, and
- Use of corrosion resistant alloys.

4.5.3 Corrosion Allowance

The selected pipe wall thickness is to include a corrosion allowance to account for internal and external corrosion during the service life of the pipe. Determination of the amounts of corrosion allowances is to account for corrosion protection methods applied, corrosion-resistant properties of the line pipe material, the fluid corrosivity inside the pipe, chemical compositions of seawater, location of the pipeline, etc. The values of the allowances are to be submitted and agreed upon between the asset designer/owner and Class. Guidance for estimating corrosion rates and allowances is given in Chap. 15. Net thickness, which means that the thickness of the corrosion allowance is deducted from the nominal wall thickness ($t_{nominal} = t_{net} + Corrosion\ allowance$), is to be used for the checks of hoop stress and hydrostatic collapse, as specified in Chap. 8. For all other load cases involving longitudinal and radial stresses, nominal wall thickness is to be used.

4.5.4   Monitoring and Maintenance of Corrosion Control Systems

Corrosion rate and the effect of anti-corrosion systems are to be evaluated by applying a monitoring programme. Remedial actions are to be taken based on the evaluation results.

Design Requirements and Loading

5

5.1 General

This chapter concerns the identification, definition and determination of loads that are to be considered in the design of pipelines. Loads generally acting on pipelines are categorised into load classes and followed by more detailed descriptions in subsequent sections, together with acceptance criteria for the utilisation of the pipe strength. The criteria cover only the plane pipe, and for flanges and other connectors, other recognised standards such as the ASME Boiler and Pressure Vessel Code are to be used. On commencement of the detailed engineering phase, a comprehensive quality plan is to be prepared, detailing the controls that will be implemented in the course of the design.

The quality plan is to set down the structure and responsibilities of the design team and outline procedures governing the assignment of design tasks, checking of work, document issues and tracking. The design process is to be fully documented and supported by comprehensive calculations in which all assumptions are fully justified. A Design Report is to be prepared in which all data analysis, calculations and recommendations are clearly laid out. Document control procedures are to ensure the traceability of all documentation, drawings, correspondence, and certification. The design steps involved are illustrated in the design flowchart in Fig. 5.1.

5.1.1 Regulations, Codes, and Standards

International codes and standards pertinent to the design, manufacture, coating, welding and inspection of pipelines and ancillary flanges and fittings are listed in Chap. 16. The Design Basis is the document that defines all the data and conditions that are required for

© The Author(s), under exclusive license to Springer Nature Switzerland AG 2025
A. A. Olsen, *Subsea Pipeline Systems*, Synthesis Lectures on Ocean Systems
Engineering, https://doi.org/10.1007/978-3-031-74790-8_5

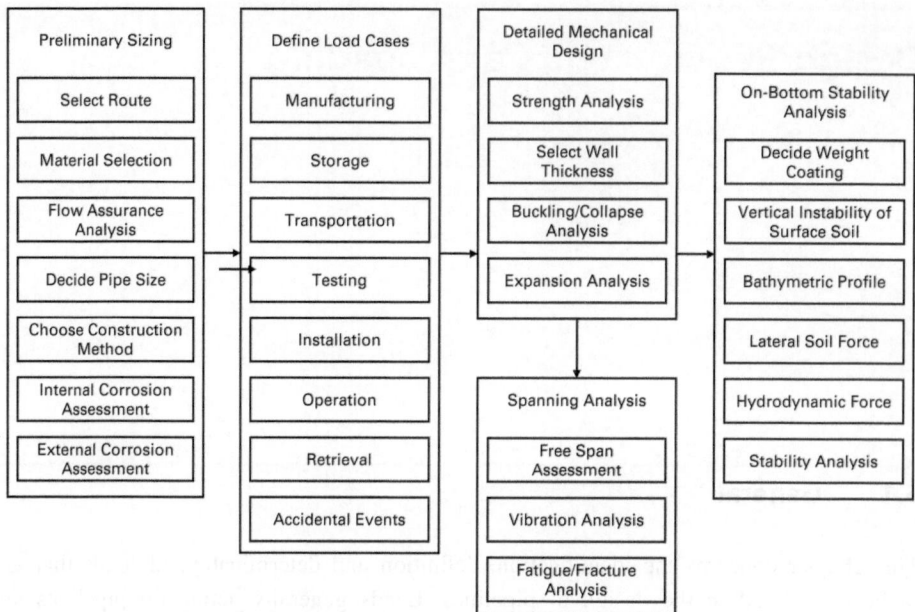

Fig. 5.1 Pipeline design flowchart

the design of the pipeline system. The document defines all codes and standards, asset owner requirements, design criteria, environmental conditions, loads and safety factors.

5.1.2   Mechanical Design

The objective of mechanical design is to determine the wall thickness and material grade for the pipeline system, determined by application of flow assurance criteria. The governing codes are to be:

- ANSI B31.4 for production and water injection pipelines,
- ANSI B31.8 for gas injection and gas export lines, and
- API RP 1111 "Design Construction, Operation, and Maintenance of Offshore Hydrocarbon Pipelines".

The mechanical design of all subsea pipelines requires compliance with API 5L standard wall thickness criteria. Mechanical design is also to ensure the on-bottom stability of each pipeline under installed and operational conditions. For each subsea pipeline, it is to be verified that the pipeline, as designed, is capable of withstanding all loads that may be reasonably anticipated over its specified design life. Imposed loads are to be classified as functional, environmental, or constructional and may be continuous or incidental, unidirectional, or cyclic in nature. Accidental loads are to be considered separately following review of risk factors for the particular development and are to be applied under agreed combinations with functional and environmental loads. All potential external and internal loads are to be identified and load combinations developed to represent superposition that may occur within defined degrees of probability. In preparing load cases, the probable duration of an event (e.g., pipeline installation) is to take account of the selection of concurrent environmental conditions.

Extreme environmental events are unlikely to coincide, and therefore, the design process is to take caution to exclude unrealistic load combinations. Pipelines are to be designed for the load combination that yields the most unfavourable conditions in terms of overall stress utilisation. Usage factors are to be determined from referenced design codes. Usage factors are to differentiate between equivalent stresses stemming from the imposition of functional and environmental loading and those derived under installation or hydrotest conditions. Should the designated code make individual provision for conveyed fluid categorisation, this is to be factored into the design equation. Due consideration is to be given to the potential effect of residual loads. Such loads are to be established, where feasible, from installation logbooks. Where an estimate of residual loading is necessary, this is to be done conservatively.

5.1.3   Load Case Definition

The subsea pipeline is to be designed to satisfy its functional requirements under loading conditions corresponding to the internal environment, external environment, system requirements and design service life. All load cases for the pipeline are to include manufacturing, storage, transportation, testing, installation, operation, retrieval, and accidental events. The design of the pipeline is to be based on design load cases, which are to be defined in a project specific Design Basis document. All load combinations and conditions are to be approved by Class.

5.1.4   Wall Thickness Selection

Pipeline design pressures are to be as defined in the design basis. Design pressures are to be calculated from flow assurance and reservoir data. The following pressures are to be considered in the design as a minimum:

• Design wellhead shut-in pressure,
• Design pressure,
• Local design pressure,

- Hydrotest pressure (at sea level),
- Leak test pressure (at sea level),
- Maximum allowable operating pressure (MAOP),
- Maximum incidental pressure,
- Maximum allowable incidental pressure, and
- Mill test pressure.

Pipeline wall thickness is to be determined on the basis of the local design pressure, defined as the internal pressure at any point in the pipeline system for the corresponding design pressure. The local pressure is to be taken as the design pressure at the reference height plus the static head of the transported medium due to the difference between the reference height and the height of the section being considered. The required pipeline wall thickness is also a function of pipeline material grade, diameter, water depth and installation methods.

5.1.5 Expansion Analysis

Analysis is to be performed to model and predict the effect of combined internal pressure and thermal gradient on the longitudinal expansion of the pipeline. Expansion analysis is to consider the most onerous load combinations to determine the greatest potential linear strain of the pipeline. Analysis is to model the surrounding soil properties, derive a representative seabed friction factor and assume a depth of burial based on the bearing properties of the soil. The ability of a pipeline to accommodate the effects of pipeline expansion is a function of the pipeline length and the virtual anchor points at which expansion forces are balanced by mobilised soil friction. Where insufficient pipeline length is available along which to mobilise frictional resistance against expansion (such as in a short, high-temperature production line), a mechanical means of absorbing pipeline movement is to be designed, such as an expansion loop or offset.

Expansion offsets are to be modelled to accurately represent the physical boundary conditions including the resistance against movement mobilised by the soil. Both pipeline and riser growth are to be modelled unless they are decoupled by the interposition of an interface structure. A finite element model is the preferred analytical method. Alternative methods may be considered upon presentation of a detailed justification of the technique proposed. Analysis of the spool piece is to consider the temperature profile along the pipeline and the pressure effects on expansion. Design of a spool-piece or offset is to consider the implications of installation. Where an alternative means of introducing pipeline flexibility is feasible, this is to be explored, such that over-complexity of the subsea pipeline system is avoided.

5.1.6 Buckle Analysis

Subsea pipelines are susceptible to cross-section instability and collapse under a combination of bending moment and external pressure. A buckle initiated by mechanical damage or external pressure may propagate along the pipeline if the external pressure is greater than the propagation pressure for the pipeline. The pipeline is to be designed to withstand local buckling under the most severe conditions of combined external pressure and bending moment that is specified for installation and operation.

A buckling analysis is to be performed to determine the wall thickness of the pipeline that will be required to resist collapse. The analysis is to determine the collapse and buckling propagation pressures in accordance with Chap. 8. The analysis is to consider the following key parameters:

- Pipeline diameter,
- Wall thickness,
- Line pipe ovality,
- Material grade and yield strength,
- Reduction in hoop compressive yield stress (caused by the manufacturing process), and
- Maximum allowable bending strain (during installation and operation).

Where necessary, limitations on ovality may be specified to increase the pipeline's resistance to buckling. The effects of plastic strains that may be imposed on the pipeline from reeled installation or from the stinger overbend during S-Lay are to be evaluated and considered in the analysis.

5.1.7 Buckle Arrestor Design

If the analysis demonstrates that the external pressure exceeds the pipeline propagation pressure, buckle arrestors are to be designed to arrest propagating buckles. A clear rationale for arrestor design is to be outlined, indicating required steps in the design and assumptions made. The use of any empirical correlations is to be fully documented and justified, indicating the limitations in validity of the data. The economics of buckle arresting is to be considered in the design to arrive at a recommended buckle arrestor spacing. Determination of arrestor spacing is to compare the consequences of failure (replacement of the buckled section of pipe plus loss of use of the line) and the cost of manufacture and installation of buckle arrestors.

5.1.8   Fracture Analysis

All pipelines are to have inherent resistance against the initiation and propagation of brittle and ductile fracture. Safety against brittle fracture is to be governed principally by material conformance requirements, as detailed in industry-recognised standards.

5.1.9   On-bottom Stability

The mechanical design of subsea pipelines is to ensure the stability of the pipeline on the seabed under extreme environmental loading conditions over the design life of the system. The pipeline is to be designed such that no movement from its installed location occurs, except by those defined limits covering lateral and vertical movement, thermal expansion, and settlement. Permissible movements are to be determined, as defined under industry-recognised standards. It is to be adequately demonstrated that permissible movements will not adversely affect pipeline integrity, neither in the short nor long term. The following aspects are to be considered in the stability analysis:

- Hydrodynamic forces due to steady-state currents, loop currents, extreme wave action, turbidity currents, river plume effects, etc.,
- Lateral soil forces,
- Potential slope instability of the seabed due to geotechnical or geo-seismic mechanisms, and
- Vertical instability of the surface soil layers.

All pipelines are to be negatively buoyant with a submerged weight sufficient to resist vertical uplift and to mobilise friction against lateral sliding. In the conduct of the design, it is to be recognised that permissible movements may vary along the length of the pipeline. The design is to adequately delineate any such zones and clearly state the assumptions made with respect to a particular portion of a pipeline's length. A dynamic method of analysis is to be used to define pipeline movements. The analysis is to use the most adverse combination of environmental conditions. Analysis is to consider directionality and a realistic variation in force incidence along the length of the pipeline. Stability analysis is to consider the pipeline under installation, testing and operational conditions.

Buried pipelines are to be constructed with adequate vertical stability such that sinking or floatation will be prevented or within a limit such that strength and stability criteria specified in Chaps. 8 and 9 are not exceeded. The soil liquefaction potential may be assessed if anticipated along the route. The trenching and naturally backfilled soft clay soil may have the tendency to liquefy due to wave action on the seabed, therefore, special attention is to be given. Environmental criteria are to be defined according to the duration of a given condition. Assessment of stability under installation conditions (including hydrotest) is to be based on a 1-year return period. Stability verification for operational conditions is to be based on 100-year return period data.

Adequate soil investigation is to be carried out to enable accurate determination of lateral and longitudinal friction coefficients. Geotechnical investigation is to establish soil characteristics such as soil type, specific weight, particle size distribution and shear strength. Additional information is to be derived from geohazard evaluation, including a probabilistic seismic assessment and mass gravity flow assessment. Interpretation of geotechnical conditions along the proposed pipeline is to identify all significant changes in characteristics liable to influence the stability assessment of the pipeline. In some cases, the use of concrete coating as a stability enhancement may not be appropriate. Should the potential for pipeline instability be identified in these cases, the following methods are to be considered:

- Increased pipe wall thickness,
- Piled anchors,
- Concrete saddles/gravity anchors,
- Flexible concrete or artificial seaweed mattresses,
- Grout bags, and
- Rock placement.

Other methods may be proposed, provided they can be demonstrated to be feasible and cost-effective alternatives. Recommendations are to be presented for approval in the course of the detailed design before the design is finalised. A full and detailed analysis of the proposed stability enhancement method is to be performed. In determining the degree of restraint, the following factors are to be considered:

- Method of load transfer between lateral restraint and pipeline, and
- Potential for flow-induced vibration between the restraint devices.

5.1.9.1 Free Span Analysis

Seabed bathymetry profiles are to be used to perform a series of free span analyses. Assuming a uniform residual pipelay tension in combination with a representative bottom roughness profile, an estimate is to be made of the magnitude and potential distribution of free spans along the pipeline route. End support conditions are to be idealised and end slope conditions are to be accounted for in the free span analysis. Spans identified through this analysis are to be tabulated, characterised by span length, support separation, maximum span clearance and static stress levels. In the water depth considered, vortex shedding will be considered to derive only from current effects. Assessment of free span under installation (including hydrotest) and operating conditions is to be based on 1-year return period and 100-year return period events, respectively. A maximum allowable span length is to be determined by application of static and dynamic criteria. In performing static analysis, it is to be permissible to consider either allowable stress or strain criteria. A maximum static stress level is to be determined by calculation such that sufficient margin is made available under the operating condition to allow for dynamic excitation.

5.1.9.2 Vortex-Induced Vibration and Fatigue

Vortex shedding may be set up in a pipeline free span due to fluid flow effects around the pipe structure, resulting in potentially severe oscillation. The aim of the design is either to prevent oscillations or to demonstrate that those oscillations are acceptable in terms of the general serviceability of the pipeline, allowable stress levels and fatigue considerations. Analysis is to calculate critical current velocities for both crossflow and in-line motion. The natural frequency of the pipeline span is a function of the pipe stiffness, end conditions, length, and effective mass. Using these parameters, the natural frequency of the pipe span is to be calculated. Calculations are to fully document the logic process used to model end pipe conditions, added mass factors, reduced velocity and stability parameter. A critical span length is to be calculated for each pipeline and the results are to be tabulated to show the limiting span in each case.

Critical span lengths are to be calculated based on the avoidance of in-line motion for the imposed design bottom current. It is to be demonstrated also that, with the selected limiting span length, crossflow motion will not occur. Fatigue evaluation is based on a Palmgren-Miner fatigue model, and the fatigue life of the pipeline is to be at least 10 times the specified service life. Prior to undertaking the fatigue analysis, an appropriate S-N curve may be proposed and utilised, provided a strong rationale for its application is given.

5.1.9.3 External Corrosion Protection Design

The subsea pipeline external corrosion is to be prevented by a combination of external anti-corrosion coating and the installation of sacrificial anodes. Sacrificial anodes are designed in accordance with the recognised codes.

5.2   Design Basis

The Design Basis is the document that defines all the data and conditions that are required for the design of the subsea pipeline. The Design Basis defines all codes and standards, the asset owner's requirements, design criteria, environmental conditions loads and safety factors that will be used for the design. Design data is to be defined in the Design Premise/ Design Basis Document. The Design Premise is to contain results of conceptual studies, environmental conditions, regulatory requirements, and applicable codes and standards. Whenever possible, the input data (and any assumptions made where data is lacking) is to be agreed with the asset owner or asset designer, and any ambiguous or conflicting requirements are to be resolved prior to the design and initiation of engineering work. As

Table 5.1 Design information for pipelines

Parameters	Remarks and observations
Internal fluid parameters	All relevant internal fluid parameters are to be specified, including parameters in Table 5.2: Internal pressure Temperature Fluid composition Service definition Fluid/flow description Flow rate parameters Thermal parameters
External environment	All relevant external parameters are to be specified, including parameters in Table 5.3.
System description	Defined by the asset owner/designer
Service life	Defined by the asset owner/designer
Design load case definitions	Defined by the asset owner/designer
Design criteria	Defined by the asset owner/designer
Analysis parameters	Defined by the asset owner/designer

a minimum, the design is to be based on the parameters specified in Tables 5.1 , 5.2, 5.3. The list is to be expanded with any additional specific parameters which may impact the design. Environmental data is to be related to months of the year or seasons. Loop current information is to be obtained for pipeline applications in areas subject to such events.

5.3   Definitions of Design Loads

Loads acting on pipelines can be divided into environmental, functional, and accidental loads.

5.3.1   Environmental Loads

Environmental loads are defined as loads imposed directly or indirectly by environmental phenomena such as waves, current, ice and snow. In general, the environmental loads vary with time and include both static and dynamic components. The characteristic parameters defining environmental loads are to be appropriate to the operational phases, such as transportation, storage, installation, testing and operation. Environmental loads and load effects are further described in Chapters 6 and 9.

Table 5.2 Internal fluid parameters

Parameters	Remarks and observations
Internal pressure	Operating pressure or pressure profile through service life Maximum design pressure Minimum design pressure Maximum allowable operating pressure
Temperature	Operating temperature or temperature profile through service life Maximum design temperature Minimum design temperature
Fluid composition	All parameters that define service conditions Corrosive agents including organic acids, chlorides, etc. Injected chemical products including corrosion, hydrate and scale inhibitors and wax solvents
Service definition	Including partial pressure of CO_2 and H_2S, and water content
Fluid/flow description	Fluid type and flow regime
Flow rate parameters	Flow rates, fluid density, viscosity
Thermal parameters	Fluid heat capacity

Table 5.3 Environment parameters

Parameters	Remarks and observations
Location	
Water depth	Design water depth, water depth variations
Seawater data	Density, pH value, electronic resistance, salinity, minimum and maximum temperatures
Air temperature	Minimum, maximum, and mean temperatures
Soil data	Description, shear strength, friction coefficients, densities
Marine growth	Maximum values and variations, marine growth density
Current data	Current as a function of water depth, direction and return period
Wave data	Significant and maximum waves, associated periods, wave spectra, scatter diagram, as a function of direction and return period
Tide level	Highest/lowest astronomical tide, storm surge
Hydrodynamic coefficients	For pipelines on the seabed or open trench

5.3.2 Functional Loads

Functional loads are defined by dead, live and deformation loads occurring during transportation, storage, installation, testing, operation, and general use.

- *Dead loads* are loads due to the weight in air of principal structures (e.g., pipes, coating, anodes, etc.), fixed/attached parts and loads due to external hydrostatic pressure and buoyancy calculated on the basis of the still water level,
- *Live loads* are loads that may change during operation, excluding environmental loads which are categorised separately. Live loads will typically be loads due to the flow, weight, pressure, and temperature of containment,
- *Deformation loads* are loads due to deformations imposed on pipelines through boundary conditions such as reel, stinger, rock berms, tie-ins, seabed contours, etc.

The functional loads are to be determined for each specific operation expected to occur during the pipeline life cycle and are to include the dynamic effects of such loads, as necessary. Load and load effects are further described in Chaps. 6 and 9. In addition, extreme values of temperatures expressed in terms of recurrence periods and associated highest and lowest values are to be used in the evaluation of pipe materials.

5.3.3 Accidental Loads

Accidental loads are defined as loads that occur accidentally due to abnormal operating conditions, technical failure, and human error. Examples are soil sliding, earthquakes, impacts from dropped objects, trawl board or collision. It is normally not necessary to combine these loads with other environmental loads unless site-specific conditions indicate such requirement. Dynamic effects are to be properly considered when applying accidental loads to the design. Risk-based analysis and previous experience may be used to identify the frequency and magnitude of accidental loads. Accidental loads and load effects are further described in Chap. 9. Typical design loads may be categorised in accordance with Table 5.4.

Table 5.4 Categorisation of design loads for pipelines

Environmental loads	Functional loads	Accidental loads
Waves	Weight in air of:	Impacts from:
Current	• Pipe	• Dropped object(s)
Tides	• Coating	• Trawl board
Surge	• Anodes	• Collision
Marine growth	• Attachments	Soil sliding
Subsidence	• Etc.	Loss of floating installation station
Scours	Buoyancy	
Seafloor instability	Towing	
Seismic	External hydrostatic pressure	
Sea ice	Internal pressures:	
Soil liquefaction	• Mill pressure test	
Hygrothermal aging	• Installation	
	• Storage, empty/water filled	
	• In-place pressure test	
	• Operation	
	Installation tension (pipes)	
	Installation bending (pipes)	
	Makeup (connectors)	
	Boundary conditions:	
	• Reel	
	• Stinger	
	• Tie ins	
	• Rock berms	
	• Seabed contours	
	• Etc.	
	Soil interaction	
	Loads due to containment:	
	• Weight	
	• Pressure	
	• Temperature	
	• Fluid flow, surge, and slug	
	• Fluid absorption	
	Inertia	
	Pigging and run tools	

Geotechnical and Environmental Conditions

6

6.1 General

The seabed stiffness and soil friction definition where the contact pressure between pipes and seabed governs the friction force can generally describe the interaction between the seabed and the pipeline. The geotechnical conditions are an important part of establishing the pipe/soil interaction model for in-place analysis, which is further discussed in Chap. 9.

6.2 Pipe Penetration in Soil

The penetration of a statically loaded pipe into soil can be calculated as a function of pipe diameter, vertical contact pressure, the undrained shear strength and submerged soil density. The circular form of pipelines leads to a combined effect of friction and bearing capacity resisting soil penetration. To represent the soft nature of the seabed, a pressure/penetration relationship can be used to specify the pipeline's penetration as a function of the ground pressure (force per unit length of the pipeline).

6.3 Soil Friction

Bi-axial soil friction data are, when applicable, to be used for in-place analysis. Both pipeline penetration of the seabed and build-up of loose sediments due to lateral movement may lead to extra lateral soil resistance. The degree of penetration/build up is dependent on the soil type and stiffness and is to be considered in the friction model.

A. A. Olsen, *Subsea Pipeline Systems*, Synthesis Lectures on Ocean Systems Engineering, https://doi.org/10.1007/978-3-031-74790-8_6

6.4   Breakout Force

The breakout force is the maximum force needed to move a pipe from its stable position on the seabed. This force can be significantly higher than the force needed to maintain the movement after breakout, depending on the soil type and degree of penetration. Reasonable breakout forces are to be used in in-place analyses.

6.5   Soil Surveys

Seabed soil is to be sampled at appropriate intervals and locations along with the bathymetrical surveys. Normally, only the upper 2 metres are of interest for pipelines, with special attention paid to the top 10 cm, where accurate data will improve the stability evaluation.

6.6   Environmental Effects

Design environmental conditions are to be defined by the operator, together with oceanographic specialists, and approved by Class. All foreseeable environmental phenomena that may influence the pipeline integrity are to be described in terms of their characteristic parameters relevant to operational and strength evaluations. Field and model-generated data are to be analysed by statistical and mathematical models to establish the range of pertinent variations of environmental conditions to be employed in the design. Methods employed in developing available data into design criteria are to be described and submitted in accordance with Chap. 1. Probabilistic methods for short-term, long-term, and extreme value predictions employing statistical distributions are to be evidenced by relevant statistical tests, confidence limits and other measures of statistical significance. Hindcasting methods and models are to be fully documented. Due to the uncertainty associated with the definition of some environmental processes, studies based on a parametric approach may be helpful in the development of design criteria. Generally, suitable environmental data and analyses will be accepted as the basis for designs when fully documented with sources, date and estimated reliability noted. For pipelines in areas where published design standards and data exist, such standards and data may be cited as reference.

6.7   Current

Current may be a major contributor to both static and dynamic loading on pipelines installed at any depth. The current velocity and direction profile at a given location may have several contributions of which the most common are:

- Oceanic scale circulation patterns,
- Lunar/astronomical tides,
- Wind and pressure differential generated storm surge, and
- River outflow.

The vector sum of all current components at specified elevations from the seafloor to the water surface describes the current velocity and direction profile for the given location. The current profile might be seasonally dependent, in which case, this is to be accounted for in the design. For pipelines, it will be enough to know the current profile associated with extreme waves at a level near the seafloor. Normally, the current velocity and direction do not change rapidly with time and may be treated as time invariant for each sea state. Onsite data collection may be required for previously unstudied areas and/or areas expected to have unusual or severe current conditions. If the current profile is not known from on-location measurements but is judged not to be severe for the design, the current velocity at a given depth may be established using a velocity profile formulation. Current velocity profiles are to be based on site specific data or recognised empirical relationships, and the worst design direction is to be assumed.

6.8   Waves

Waves are a major source of dynamic loads acting on pipelines located in shallow waters (normally <150 m), and their description is therefore of high importance. Statistical site-specific wave data, from which design parameters are to be determined, are normally to include the frequency of occurrence for various wave height groups and associated wave periods and directions. For areas where prior knowledge of oceanographic conditions is insufficient, the development of wave dependent design parameters is to be performed in cooperation with experienced specialists in the fields of meteorology, oceanography, and hydrodynamics. For a fully developed sea, waves may be represented using the Bretschneider spectrum while the JONSWAP spectrum normally will be applicable for less developed seas. In the calculation of spectrum moments, a proper cut-off frequency based on a project-defined confidence level is to be applied. Wave scatter diagrams can be applied to describe the joint probability of occurrence of the significant wave height and the mean zero-crossing period. Where appropriate, alternative traditional regular wave approaches may be used.

When dealing with extreme response estimations, the regular design wave heights are to be based on the maximum wave height of a given return period, e.g., 1, 10 or 100 years, found from long-term wave statistics. The estimation of the corresponding extreme wave period is, in general, more uncertain due to lack of reliable data, and it is consequently recommended that the wave period be varied over a realistic interval to ensure that all extreme wave cases have been considered. For systems with obvious unfavourable wavelengths and periods due to geometry or eigen-frequencies, the design wave period can be identified based on such criteria while the wave height follows from breaking wave criteria or statistical considerations. Long-term response statistics are to be applied in fatigue damage assessment whereby a scatter diagram of the joint probability of the sea state vector and the wave spectrum represents the wave climate defined by significant wave height, peak period, and main wave direction. A simplified representation of the long-term distribution for the response may be based on the frequency domain method consisting of:

- Establishing an approximate long-term response distribution based on stochastic dynamic analyses,
- Calculation of an approximate lifetime extreme response,
- Identification of the design storm, and
- Estimation of lifetime maximum response based on time domain simulations.

In analysis, a sufficient range of realistic wave periods and wave crest positions relative to pipelines are to be investigated to ensure an accurate determination of the maximum wave loads. Consideration is also to be given to other wave-induced effects such as wave impact loads, dynamic amplification, and fatigue. The need for analysis of these effects is to be assessed on the basis of the configuration and behavioural characteristics of pipelines, the wave climate and past experience.

6.9   Combinations of Current and Waves

The worst combination of current and waves is to be addressed in the design. When current and waves are superimposed, the current velocity and direction are to be added as vectors to the wave-induced particle velocity and direction prior to computation of the total force, and where appropriate, flutter and dynamic amplification due to vortex shedding are to be account for. For pipelines having small diameters compared to the wave lengths being considered, semi-empirical formulations such as Morison's equation are considered to be an acceptable basis for determining the hydrodynamic force acting on a pipe:

$$F = F_D + F_i$$

where:

F = in-line component of hydrodynamic force per unit length along pipes
F_D = hydrodynamic drag force per unit length
F_i = hydrodynamic inertia force per unit length

The drag force for a stationary pipe is given by:

$$F_D = \frac{1}{2}\rho \cdot OD \cdot C_D \cdot u_n \cdot |u_n|$$

where:

ρ = density of water
OD = total external diameter of pipe, including coating, etc.
C_D = drag coefficient
u_n = component of the total fluid velocity vector normal to the axis of pipes

The inertia force for a stationary pipe is given by:

$$F_i = \rho \cdot \left(\frac{\pi \cdot OD^2}{4}\right) \cdot C_M \cdot a_n$$

where:

C_M = inertia coefficient based on the displaced mass of fluid per unit length
a_n = component of the total fluid acceleration vector normal to the axis of pipes

The lift force for a stationary pipe located on or close to the seabed is given by:

$$F_L = \frac{1}{2}\rho \cdot C_L \cdot u_n^2 \cdot OD$$

where:

F_L = lift force per unit length
C_L = lift coefficient

For pipelines located on or close to the seabed, the hydrodynamic force coefficients C_D, C_M and C_L will vary as a function of the gap.

6.10   Tides

Tides, when relevant, are to be considered in the design of pipelines. Tides may be classified as lunar or astronomical tides, wind tides and pressure differential tides. The combination of the latter two is defined as "storm surge" and the combination of all three as "storm tide". The water depth at any location consists of the mean depth, defined as the vertical distance between the seabed and an appropriate near-surface datum and a fluctuating component due to astronomical tides and storm surges. The highest and the lowest astronomical tide bound the astronomical tide variation. Still water level is to be taken as the sum of the highest astronomical level plus the storm surge. Storm surge is to be estimated from available statistics or by mathematical storm surge modelling.

6.11   Marine Growth

Marine growth may accumulate and is to be considered in the design of pipelines. The highest concentrations of marine growth will generally be seen near the mean water level with an upper bound given by the variation of the daily astronomical tide and a lower bound, dependent on location. Estimates of the rate and extent of marine growth may be based on previous experience and available field data. Particular attention is to be paid to increases in hydrodynamic loading due to the change of:

• External pipe diameter,
• Surface roughness,
• Inertial mass, and
• Added weight.

Consideration is also to be given to the fouling effects likely on corrosion protection coatings.

6.12   Subsidence

The effects of seafloor subsidence are to be considered in the overall design when pipelines are installed in areas where unique geological conditions exist. This will be an area where, for example, significant seafloor subsidence could be expected to occur as resulting from depletion of the subsurface reservoir. The magnitude and time scale of subsidence are in such cases to be estimated based on geologic studies.

6.13   Scours

The seafloor contours in installation areas may change considerably over time due to scour erosion, which is removal of soil due to current and waves caused either by natural geological processes or by structural elements interrupting the flow regime near the seafloor. When applicable, the magnitude and time scale of scour erosion is to be estimated based on geologic studies and its impact on design appropriately accounted for. When the magnitude of scour erosion makes it difficult to account for in design, a proper survey programme and routines for evaluating observed seafloor changes is to be established.

6.14   Seafloor Instability

Seafloor instability may be seen under negligible slope angles in areas with weak, under-consolidated sediments. Movements of the seafloor may be activated resulting from loads imposed on the soil due to pipeline installation, change in pipeline operating conditions, wave pressure, soil self-weight, earthquakes or combinations of these phenomena. When applicable, such areas are to be localised by proper surveys and precautions such as rerouting taken in the design.

6.15   Seismic

The seismic activity level for the pipeline installation area is to be evaluated based on previous records or detailed geological investigations. For pipelines located in areas that are considered seismically active, the effects of earthquakes are to be considered in the design. An earthquake of magnitude that has a reasonable likelihood of not being exceeded during the design life is to be used to determine the risk of damage, and a rare intense earthquake is to be used to evaluate the risk of structural failure. These earthquake events are referred to as the Strength Level and Ductility Level earthquakes, respectively. The magnitudes of the parameters characterising these earthquakes, having recurrence periods appropriate to the design life of the pipelines, are to be determined. The effects of earthquakes are to be accounted for in design, but generally need not be taken in combination with other environmental factors.

The strength level and ductility level earthquake-induced ground motions are to be determined on the basis of seismic data applicable to the installation location. Earthquake ground motions are to be described by either applicable ground motion records or response spectra consistent with the recurrence period appropriate to the design life of pipelines. Available standardised spectra applicable to the region of the installation site are acceptable, provided such spectra reflect site-specific conditions affecting frequency content, energy distribution and duration. These conditions include the type of active faults in the region, the proximity to the potential source faults, the attenuation or amplification of ground motion and the soil conditions.

The ground motion description used in design is to consist of three components corresponding to two orthogonal horizontal directions and the vertical direction. All three components are to be applied to pipelines simultaneously. As appropriate, effects of soil liquefaction, shear failure of soft mud and loads due to acceleration of the hydrodynamic added mass by the earthquake, mud slide, tsunami waves and earthquake-generated acoustic shock waves are to be accounted for in the design.

6.16   Sea Ice

Arctic pipelines in shallow water can be damaged by grounded ice which drags along the seabed and cuts gouges that can be several meters deep. The intensity of gouging depends on the interaction between the ice climate, wind, water depth, the local topography of the bottom and the seabed geotechnics. Severe deformation of the seabed extends below the bottom of the gouge so that a pipeline can be damaged even though the ice passes above it. More details about conditions that are to be addressed in design and construction for arctic and subarctic offshore regions can be found in API RP 2N.

6.17   Environmental Design Conditions

In this guide, the combination of environmental factors producing the most unfavourable effects on pipelines as a whole, and as defined by the parameters given above, is referred to as the environmental design conditions. The combination and severity of environmental conditions for use in design are to be appropriate to the pipelines and consistent with the probability of simultaneous occurrence of the environmental phenomena. It is to be assumed that environmental phenomena may approach pipelines from any direction unless reliable site-specific data indicate otherwise. The direction, or combination of directions, which produces the most unfavourable effects on pipelines is to be accounted for in the design, unless there is a reliable correlation between directionality and environmental phenomena. When applicable, it is recommended that at least the following environmental conditions be covered by pipeline analyses.

Normal operating condition with pipeline in the normal intact status	• Worst combination of 1 year wind, wave, current and tide • Environmental condition of 100-year return waves plus 10-year return current • Environmental condition of 10-year return waves plus 100-year-return current • Realistic values for marine growth • Realistic values for loads due to sea ice, and • Realistic values for earthquakes.

(continued)

(continued)

Temporary condition The following are to be checked as temporary conditions:	• *Transportation condition* Geometrical imperfections such as dents, and out-of-roundness introduced by loads applied during transportation are to be considered • *Installation/retrieval condition* Varying amount deployed Filled with air or water Environmental condition of 1-year wave and current or reliable weather forecasts • *System pressure test* Loads (especially pressure loads) during system pressure test • *Shut-down and start-up* The fatigue evaluation is to include loads induced by shutdown and startup • *Pigging condition* Loads induced by pigging operations are to be considered
Abnormal/accidental operating condition Impacts and soil sliding conditions are to be checked, when applicable	
Fatigue Adequate loading conditions are to be used for fatigue load effect analysis, including:	• Significant wave • Vortex–induced vibrations • Cyclic loading induced during installation and operation, and • Thermal stresses induced during processing and operation

Flow Assurance Analysis

7

7.1 General

Flow assurance analysis is not generally subject to Class review. However, reports for a full thermal–hydraulic analysis may be required by Class as supporting documents. The thermal–hydraulic analysis is to determine the optimum size of the subsea pipelines and to predict pipeline temperature and pressure profiles, flow regimes and liquid holdups in steady state and transient conditions.

7.2 Pipeline Sizing/Steady State Analysis

Sizing of subsea pipelines is to be carried out with the aid of hydraulic analysis. Initial sizing may be performed using hand calculations or a suitable computer programme, but a validated computer programme appropriate to the type of fluid flow and conditions being considered is always to confirm final sizing. Although hydraulic considerations usually decide the size of a pipeline and associated connected systems, several non-hydraulic criteria may have an effect and are to be considered. Some of the criteria are:

- Riser design (the requirement for constant inner diameter),
- Standard pipe sizes,
- Fluid velocity control, and
- Piggability.

© The Author(s), under exclusive license to Springer Nature Switzerland AG 2025　　　57
A. A. Olsen, *Subsea Pipeline Systems*, Synthesis Lectures on Ocean Systems
Engineering, https://doi.org/10.1007/978-3-031-74790-8_7

7.3 Fluid Phase Definition

Fluid phase behaviour is to be determined from the flow analysis and phase envelopes are to be developed for the produced fluids. The information is to be used for hydrate and wax deposition analyses.

7.4 Pressure and Temperature Profile

Pressure and temperature profiles are to be established for the single phase (gas or liquid) and multiphase risers. The prediction of flow rate and pressure drop in multi-phase pipelines is not as accurate as for the single-phase lines and the same applies to the prediction of liquid holdup and slug sizes. The sensitivity of the pipeline diameter selection is to be investigated, and consideration is to be given to selection of a smaller or larger diameter to accommodate the uncertainties in the calculations.

7.5 Transient Analysis

A variety of transient thermal–hydraulic analyses may be needed for the complete design and optimisation of a production system. These include:

- Riser and pipeline warm-up and cool-down,
- Startup and shutdown fluid flow rates and flow stabilisation time,
- Pigging, and
- Slugging.

7.6 Surge Analysis

A surge analysis may be required to predict the behaviour of the flowing fluids if slugs are present in the line and during emergency shutdowns.

7.7 Heat Transfer/Insulation Design

The subsea pipeline will require flow path insulation to retain fluid temperature and/or to increase cooldown times during short-term shut-in conditions. The insulation may be solid material, foam, or syntactic foam. The insulation material properties are to be reviewed against hydrostatic collapse. The heat transfer calculations are to take into consideration reduction of insulating properties due to reduction of insulating layer thickness (caused by hydrostatic pressure, installation loads, etc.), absorption of water and aging.

7.8 Hydrate Mitigation

Hydrate formation is a potential problem in pipelines that contain gas and free water. System design must ensure that the conditions in the riser cannot lead to hydrate formation. This is normally achieved by either keeping the pipeline warm and/or by injecting methanol or glycol in the riser/pipeline system. The design is also to cater for shutdown conditions when considering hydrate formation.

7.9 Wax Deposition

Some crude oils contain significant proportions of paraffin compounds which will start to crystallise if the temperature of the oil drops sufficiently. This can result in wax deposition on the walls of the pipeline and may result in restricted flow or blockage. System design is to ensure that the required throughput of the pipeline system can be maintained and that it can be restarted after a shutdown. Several methods of wax deposit mitigation must be considered, including insulation and/or burial to maintain higher temperature, and or the use of chemical additives to reduce wax deposition or pressure loss in the pipeline system.

Strength and Stability Criteria

8

8.1 General

This chapter defines strength and stability criteria which are to be applied as limits for the design of subsea metallic pipelines. The criteria are applicable for wall-thickness design as well as installation and in-place analyses. Alternative strength criteria based on recognised codes/standards, mechanical tests, or advanced analysis methods such as those outlined in Chaps. 14, 15 and 16 may be applied in the design based on approval by Class. If alternative strength criteria are applied in the design, consistency is to be maintained. The strength criteria listed in this chapter follow a working stress design approach and cover the following failure modes:

- Yielding,
- Local buckling,
- Global buckling,
- Fatigue, and
- Cross sectional out-of-roundness.

8.2 Stress Criteria for Metallic Gas Transportation Pipelines

8.2.1 Hoop Stress Criteria

In selecting the wall-thickness of gas transportation pipe, consideration is to be given to pipe structural integrity and stability during installation and operation, including pressure containment, local buckling/collapse, global buckling, on-bottom stability, protection

A. A. Olsen, *Subsea Pipeline Systems*, Synthesis Lectures on Ocean Systems
Engineering, https://doi.org/10.1007/978-3-031-74790-8_8

against impact loads and free span fatigue, as well as high temperature and uneven seabed-induced loads. The internal pressure containment requirements often used as a basis for wall-thickness design are:

$$\sigma_h \leq \eta_h \cdot SMYS \cdot k_T$$

where:

σ_h	hoop stress
SMYS	Specified Minimum Yield Strength of the material
k_T	temperature dependent material strength de-rating factor, as specified in Table 841.116A of ASME B31.8
ηh	0.72, hoop stress usage factor

The hoop stress σ_h for pipes is to be determined by:

$$\sigma_h = \frac{(p_i - p_e) \cdot (D - t)}{2 \cdot t}$$

where:
p_i internal design pressure
p_e external design pressure
D nominal outside steel diameter of pipe
t net wall thickness

For thick-walled pipes with a $D/t < 20$, the above hoop stress criteria may be adjusted based on BSI BS 8010-3, for example.

8.2.2 Longitudinal Stress

To ensure structural integrity against longitudinal forces, the following longitudinal stress criteria are to be satisfied:

$$\sigma_\ell \leq \eta_\ell \cdot SMYS$$

where:

σ_ℓ	longitudinal stress
SMYS	Specified Minimum Yield Strength of the material
η_ℓ	0.8, longitudinal stress usage factor

8.2.3 Von Mises Combined Stress

The von Mises combined stress, also referred to as the stress intensity, is at any point in the pipe to satisfy the following:

$$\sigma_e = \sqrt{\sigma_\ell^2 + \sigma_h^2 - \sigma_\ell \cdot \sigma_h + 3 \cdot \sigma_{\ell h}^2} \leq \eta_e \cdot SMYS$$

where:

σ_e Von Mises combined stress
σ_ℓ longitudinal normal stress
σ_h hoop stress (normal stress circumference direction)
$\sigma_{\ell h}$ shear stress due to shear force and torsional moment
η_e 0.9, usage factor for combined stress

   Note: Nominal pipe wall thickness, as specified in Chap. 4, is to be used in the calculations of the principal stresses σ_ℓ, σ_h, and $\sigma_{\ell h}$.

8.3   Stress Criteria for Metallic Liquid Transportation Pipelines

8.3.1 Hoop Stress Criteria

In selecting the wall-thickness of liquid transportation pipe, consideration is to be given to pipe structural integrity and stability during installation and operation, including pressure containment, local buckling/collapse, global buckling, on-bottom stability, protection against impact loads and free span fatigue, as well as high temperature and uneven seabed induced loads. The internal pressure containment requirements often used as a basis for wall-thickness design are:

$$\sigma_h \leq \eta_h \cdot SMYS$$

where:

σ_h hoop stress
$SMYS$ Specified Minimum Yield Strength of the material
η_h 0.72, hoop stress usage factor

The hoop stress σ_h for pipes is to be determined by:

$$\sigma_h = \frac{(p_i - p_e) \cdot (D - t)}{2 \cdot t}$$

where:

p_i internal or external design pressure
p_e external design pressure
D nominal outside steel diameter of pipe
t net wall thickness

For thick-walled pipes with a $D/t < 20$, the above hoop stress criteria may be adjusted based on, e.g., BSI BS 8010-3.

8.3.2   Longitudinal Stress

To ensure structural integrity against longitudinal forces, the following longitudinal stress criteria are to be satisfied:

$$\sigma_\ell \leq \eta_\ell \cdot SMYS$$

where:

σ_ℓ longitudinal stress
$SMYS$ Specified Minimum Yield Strength of the material
η_ℓ 0.8, longitudinal stress usage factor

8.3.3   Von Mises Combined Stress

The combined stress, also referred to as the stress intensity, is at any point in the pipe to satisfy the following:

$$\sigma_e = \sqrt{\sigma_\ell^2 + \sigma_h^2 - \sigma_\ell \cdot \sigma_h + 3 \cdot \sigma_{\ell h}^2} \leq \eta_e \cdot SMYS$$

where:

σ_e Von Mises combined stress
σ_ℓ longitudinal normal stress
σ_h hoop stress (normal stress circumference direction)
$\sigma_{\ell h}$ shear stress due to shear force and torsional moment
η_e 0.9, usage factor for combined stress

   Note: nominal pipe wall thickness, as specified in Chap. 4, is to be used in the calculations of the principal stresses σ_ℓ, σ_h, and $\sigma_{\ell h}$.

8.4 Global Buckling

Internal overpressure and increased operating temperatures may aggravate build-up of compressive forces in a pipeline which after start-up or after repeated start-up/shut-down cycles, may lead to global buckling of the pipeline. This effect is to be explicitly dealt with in the design, either by advanced analysis predicting the position and amplification of buckles or by demonstrating that the build-up of compressive force is less than the force needed to initiate global buckling.

8.5 Local Buckling/Collapse for Metallic Pipelines

8.5.1 Collapse Under External Pressure

The characteristic buckling pressure can be calculated based on:

$$p_c = \frac{p_{el}p_p}{\sqrt{p_{el}^2 + p_p^2}}$$

where:

p_{el} $\frac{2 \cdot E}{1-v^2} \cdot \left(\frac{t}{D}\right)^3$ elastic buckling pressure

p_p $SMYS \cdot \frac{2 \cdot t}{D}$ yield pressure at collapse

$SMYS$ Specified Minimum Yield Strength

E Young's Modulus

v Poisson's ratio, 0.3 for steel pipelines

In the calculation of elastic buckling pressure (p_{el}), the wall thickness is to be the net thickness of the pipe wall, as specified in Chap. 4. The pipeline is not considered to collapse if the minimum differential pressure on the pipe satisfies the following:

$$(p_e - p_i) \le \eta_b p_c$$

where:

p_e external pressure

p_i internal pressure

η_b buckling usage factor

       0.7   for seamless or ERW pipe

       0.6   for cold expanded pipe

8.5.2 Local Buckling/Collapse Under External Pressure and Bending

For installation and temporary conditions where the pipe may be subjected to external overpressure, cross-sectional instability in the form of local buckling/collapse is to be checked. For pipes with a D/t less than 50 and subjected to external overpressure combined with bending, the following strain check is to be applied:

$$\frac{\varepsilon}{\varepsilon_b} + \frac{p_e - p_i}{p_c} \leq g(f_0)$$

where:

ε	bending strain in the pipe
ε_b	$\frac{t}{2D}$, buckling strain under pure bending
p_e	external pressure
p_i	internal pressure
f_0	out-of-roundness, $(D_{max} - D_{min})/D$, not to be taken less than 0.5%
$g(f_0)$	$(1 + 10 f_0)^{-1}$, out-of-roundness reduction factor

An out-of-roundness higher than 3% is not allowed in the pipe without further analysis considering collapse under combined loads, propagating buckling and serviceability of the pipe.

8.6 Propagating Buckles

During installation or, in rare situations, shutdown of pipelines, local buckles/collapse may start propagating along the pipe with extreme speed driven by the hydrostatic pressure of seawater. Buckle arrestors may be used to stop such propagating buckles by confining a buckle/collapse failure to the interval between arrestors. Buckle arrestors may be designed as devices attached to or welded to the pipe or they may be joints of thicker pipe. Buckle arrestors will normally be spaced at suitable intervals along the pipeline for water depths where the external pressure exceeds the propagating pressure level.

Buckle arrestors are to be used when:

$$p_e - p_i \geq 0.72 \cdot p_{pr}$$

where:

p_{pr} $6 \cdot SMYS \cdot \left(\frac{2 \cdot t}{D}\right)^{2.5}$ buckle propagation pressure.

When required, buckle arrestors are to be designed according to recognised codes, such as API RP 1111.

8.7   Fatigue for Metallic Pipelines

Pipelines may be subjected to fatigue damage throughout their entire life cycle. The main causes of fatigue failure are normally consequences of:

- Installation,
- Startup and shutdown cycles, and
- Wave and current conditions.

The fatigue life may be predicted using an S–N curve approach and Palmgren–Miner's rule. The fatigue life is not to be less than ten (10) times the service life for the pipeline. This implies for the fatigue equations listed in this guide that the maximum allowable damage ratio η is not to be taken higher than 0.1. A value higher than 0.1 may be accepted if documentation for inspection, improvement in welding, workmanship is provided and accepted by Class. Typical steps required for fatigue analysis using the S–N approach are outlined below:

(1) Estimate long-term stress range distribution,
(2) Select appropriate S–N curve,
(3) Determine stress concentration factor, and
(4) Estimate accumulated fatigue damage using Palmgren–Miner's rule:

$$D_{fat} = \sum_{i=1}^{M_c} \frac{n_i}{N_i} \leq \eta$$

where:

D_{fat} accumulated fatigue damage
η usage factor for allowable damage ratio
N_i number of cycles to failure at the i-th stress range defined by the S–N curve
n_i number of stress cycles with stress range in block i

The maximum allowable damage ratio may be relaxed if detailed analysis based on fracture mechanics or reliability-based calibration demonstrates that the target safety level is fulfilled with a higher allowable damage ratio. Also, documentation demonstrating detailed fatigue inspection and planning may lead to the acceptance of a higher allowable damage ratio. The use of and documentation for using a higher allowable damage ratio is to be submitted to and approved by Class. Fatigue assessment may be based on nominal stress or hot-spot stress. When the hot spot stress approach is selected, stress concentration factors due to misalignment (for example), are to be estimated using appropriate stress analysis or stress concentration factor equations.

8.8   Out-of-Roundness

Out-of-roundness in this guide is defined as $(D_{max} - D_{min})/D$. Out-of-roundness may occur in the manufacturing phase, during transportation, storage, installation, and operation. Out-of-roundness may aggravate local buckling or complicate pigging and is to be considered. If excessive cyclic loads are expected during installation and operation, it is recommended that out-of-roundness due to through-life cyclic loads be simulated, if applicable.

8.9   Allowable Stresses for Supports and Restraints

Maximum allowable shear and bearing stresses in structural supports and restraints are to follow applicable Class Rules, AISC ASD Manual of Steel Construction, API RP 2A-WSD or alternatively recognised Rules or standards.

8.10   Installation

During installation when the pipe is fully supported on the lay vessel, relaxed criteria in the form of maximum allowable strain, e.g., may be applied when documentation for the criteria is submitted and approved by Class.

Pipeline Rectification and Intervention Design

9

9.1 General

This chapter covers on-bottom stability and global buckling requirements for seabed intervention design related to pipes. The basic seabed intervention design may be performed using simple design equations, while more challenging design scenarios often will require a more advanced approach, such as the finite element method.

9.2 In-Place Analysis

The in-place model is to be able to analyse the in-situ behaviour of a pipeline over the through-life load history. The through-life load history consists of several sequential load cases, such as:

- Installation,
- Testing (water filling and system pressure test),
- Operation (content filling, design pressure and temperature),
- Shut down and cool down cycles, and
- Storage.

9.2.1 Parameters and Procedures for In-Place Analysis

When performing in-place analyses, static and dynamic loads are to be applied, as applicable. Static analyses are to be performed, when applicable, to handle nonlinearity from large-displacement effects, material nonlinearity and boundary nonlinearity, such as contact, sliding and friction (pipe/seabed interaction). Dynamic analyses may be used in the study of nonlinear dynamic responses of pipes. General nonlinear dynamic analysis uses implicit integration of the entire model to calculate the transient dynamic response of the system. Implicit time integration is to be applied where a set of simultaneous nonlinear dynamic equilibrium equations is to be solved at each time increment.

When applicable, geometrical nonlinearity is to be accounted for in the model. The instantaneous (deformed) state of the structure is to be updated through the analysis. This is especially important when performing dynamic analysis of pipes subjected to wave loading. By including geometrical nonlinearity in the calculation, a finite element programme will use the instantaneous coordinates (instead of the initial) of the load integration points on the pipe elements when calculating water particle velocity and acceleration. This ensures that even if some parts of the pipeline undergo extreme lateral displacements, the correct drag and inertia forces will be calculated on each of the individual pipe elements that make up the pipeline.

9.2.2 Modal Analysis

The aim of the modal analysis is to calculate the natural frequencies and corresponding mode shapes. To obtain natural frequencies, modal shapes and associated normalised stress ranges for the possible modes of vibrations, a dedicated finite element programme is to be used, and as a minimum, the following aspects are to be considered:

(1) Flexural behaviour of the pipeline is modelled, considering the bending stiffness and the effect of axial force,
(2) Effective axial force that governs the bending behaviour of the span is to be accounted for, and
(3) Interaction between a spanning pipe section and pipe lying on the seabed adjacent to the span is to be considered.

Due consideration to points (1) and (2) above is to be given in both single-span and multiple-span modal analyses. The axial effective force, i.e., the sum of the external forces acting on the pipe, is also to be accounted for. It is noted that the effective force may change considerably during the various phases of the design life. It is important to ensure that a realistic load history is modelled prior to performing the modal analyses.

9.2.3   Location Fixity

The pipeline is to be designed with a specified tolerance of movement from its as-installed position. An analysis to determine that the anchoring arrangements, soil strength and friction coefficient will limit the pipeline movement to the specified tolerance is to be performed. The types of pipeline movements to be considered include horizontal movement caused by current and wave forces, vertical movement caused by hydrodynamic lift on the pipeline, either positive or negative vertical movements caused by loss of strength of the supporting or overburden soil and earthquake-induced effects.

9.2.4   High Pressure/High Temperature

Design of high pressure/high temperature pipes is to consider global buckling due to thermal and internal pressure induced expansion.

9.3   On-Bottom Stability

Unless accounted for in the design, pipelines resting on the seabed, trenched, or buried, are not to move from their as-installed position under even extreme environmental loads. The lateral stability of pipelines may be assessed in design using two-dimensional static or three-dimensional dynamic analysis methods. The submerged weight of the pipe is to be established based on the on-bottom stability calculations, which will have a direct impact on the required pipe-lay tensions, installation stresses and pipe configuration on the sea-bottom. Properly performed stability analysis, performed in accordance with, e.g., AGA L51698, will in general be acceptable to Class.

9.3.1   Static Stability Criteria

Lateral stability criteria employ a simplified stability analysis based on a quasi-static balance of forces acting on the pipe. The lateral stability analysis method may be defined as below:

$$\gamma(F_D - F_i) \leq \mu(W_{sub} - F_L)$$

where:

F_D hydrodynamic drag force per unit length
F_i hydrodynamic inertia force per unit length
F_L hydrodynamic lift force per unit length

W_{sub} submerged pipe weight per unit length
μ lateral solid friction coefficient
γ safety factor, equal to or larger than 1.1

The effect of seabed contours is, if applicable, to be included in the analysis. Buried pipes are to have adequate safety against sinking or flotation. Sinking is to be checked, assuming that the pipe is water filled, and flotation is to be checked, assuming that the pipe is gas or air filled. If the pipe is installed in soils having low shear strength, the soil bearing capacity is to be evaluated. Axial stability due to thermal and pressure effects is to be checked through simplified methods or a detailed in-place analysis. The axial stability criteria may be defined as those for lateral stability criteria. The anodes are to be acceptable to sustain the anticipated axial friction force. Axial stability is to be evaluated using suitable soil/pipe interaction models. For weight calculations of corroded pipelines, the expected average weight reduction due to metal loss is to be deducted. Reduction of soil shear strength, e.g., due to hydrodynamic loads, is to be considered in the on-bottom stability design.

9.3.2 Dynamic Stability Analysis

Dynamic analysis involves full dynamic simulation of the pipeline resting on the seabed, including modelling of soil resistance, hydrodynamic forces, boundary conditions and dynamic response. It may be used for detailed analysis of critical areas along a pipeline, such as pipeline crossings, etc., where a high level of detail is required. The acceptance criteria for dynamic analysis are defined based on strength criteria, deterioration/wear of coating, geometrical limitations of supports and distance from other structures, etc. The allowable lateral displacement of pipelines is to be based on factors such as:

- Distance from platform or other constraint,
- Seabed obstructions,
- Width of surveyed corridor,
- Significant damage to external coating or anodes due to movement,
- Interference with other pipelines or subsea installations due to movement,
- Change in load condition due to movement, and
- Change in seabed features in adjacent areas.

9.4   Free Spanning Pipeline

This section gives acceptance criteria for vortex-induced vibrations of free spanning pipes and describes a methodology applicable for design of pipeline systems. Free spans may form resulting from of the seabed topography or may be formed subsequently resulting from soil erosion and transportation. Where geotechnical and bottom-profiling survey techniques identify areas prone to erosion/deposition action, such areas are to be flagged as regions of potential pipeline span formation. The development of a free-span rectification methodology is to account for the local soil bearing strengths. For regions of low bearing strength, a mattress-type foundation may be appropriate. Alignment sheets are to be prepared indicating bathymetry, route corridor and pipeline details. A centreline profile is to be developed to show bathymetry along with core sample locations and corresponding interpreted geological cross-section. Scales and contour intervals are to be specified and alignment sheets are to indicate all existing facilities, pipelines, cables, wrecks, major obstructions, debris, and all other pertinent features, natural or man-made.

Seabed bathymetry profiles are to be used to perform a series of free span analyses. Assuming a uniform residual pipelay tension in combination with a representative bottom roughness profile, an estimate is to be made of the magnitude and potential distribution of free spans along the pipeline route. End support conditions are to be idealised and end slope conditions are to be accounted for in the free span analysis. Spans identified through this analysis are to be tabulated, characterised by span length, support separation, maximum span clearance and static stress levels. A maximum allowable span length is to be determined by application of static and dynamic criteria. In performing static analysis, it may be permissible to consider either allowable stress or strain criteria. A maximum static stress level is to be determined by calculations such that sufficient margin is made available under the operating condition to allow for dynamic excitation. Vortex shedding may be set up in a pipeline free span due to fluid flow effects around the pipe structure, resulting in potentially severe oscillation. The aim of the design is to be either to prevent oscillations or to demonstrate that oscillations are acceptable in terms of the general serviceability of the pipeline, allowable stress levels and fatigue considerations. Analysis is to obtain critical velocities for both crossflow and in-line motion. Design criteria applicable to different environmental conditions may be defined as follows:

(1) Peak stresses or strain under extreme loading conditions are to satisfy the strength criteria given in Chap. 8.
(2) Cyclic stress ranges smaller than the cut-off stress may be ignored in fatigue analyses.
(3) The allowable fatigue damage (refer to Chap. 8) is not to be exceeded.

9.4.1   Evaluation of Free Spanning Pipelines

When applicable, single span analysis is to be performed to assess the onset of in-line and cross- flow vortex-induced vibrations (VIV), as well as to calculate fatigue damage. For a cylindrical pipe in water, vortex-shedding frequency can be calculated as:

$$f_s = \frac{St \cdot u_n}{D}$$

where:

f_s vortex-shedding frequency
u_n velocity of water normal to the pipelines
D outside diameter of pipe
St Strouhal number, varies from 0.2 to 0.4 in most practical cases, but it is also a function of the Reynolds number

The natural frequency of the pipeline span is a function of the pipe stiffness, end conditions, length and added and effective mass. The natural frequency of the pipe span is to be calculated using these parameters. Calculations are to fully document the logic process used to model end pipe conditions, added mass factors, reduced velocity and stability parameter. A critical span length is to be calculated for each pipeline and the results tabulated to show the limiting span in each case. A modal analysis of a single span with appropriate boundary conditions may be conducted to calculate the natural frequencies and modes for vortex-induced vibration assessment. Different boundary conditions are to be analysed together with a range of axial forces. The natural frequency of the span may be calculated as:

$$f_n = \frac{CK^2}{2\pi L^2} \sqrt{\frac{E1}{m} \left(1 + \frac{P}{P_E}\right)}$$

where:

f_n natural frequency, in cycles per second
C coefficient, 0.7 for pipes in water and 1.0 for pipes in air
K end-fixity condition constant,
    3.14   for hinged-hinged condition.
    3.92   for fixed-hinged condition.
    4.73   for fixed–fixed condition.
I moment of inertia of pipe account for weight coating, etc.
L span length
P the effective axial force (tension positive)
f_s the Euler buckling force

$\pi^2 EI/L^2$ for hinged-hinged condition.

$2\pi^2 EI/L^2$ for fixed-hinged condition.

$4\pi^2 EI/L^2$ for fixed–fixed condition.

m mass per unit length of pipe plus mass of water displaced by pipe, internal fluid, and weight coating, etc.

For a large range of current velocity, vortex-shedding frequency is locked-in at the natural frequency of the pipe. Large amplitude vibrations may occur unless the natural frequency is sufficiently greater than the vortex-shedding frequency. A multiple span analysis may be conducted to account for the interaction between adjacent vertical spans, i.e., several spans may respond as a system. The multiple span approach may be necessary for the vertical mode of vibration where the seabed between adjacent spans form a fixed point about which the pipeline pivots during the vibration. Alternative methods, such as finite element analysis coupled with computational fluid dynamics (CFD), can be employed to accurately estimate structural response due to vortex shedding. For pipes installed in shallow waters, wave-induced in-line fatigue is to be evaluated. The in-line motion for a free spanning pipe subjected to wave forces represented by the Morison force, damping forces and axial forces may be determined using nonlinear partial differential equations. The pipe motion as a function of time and position may be obtained by solving the equation of in-line motion using modal analysis.

9.4.2 Span Correction

Critical span lengths are to be calculated based on the avoidance of in-line motion for the imposed design bottom current. It is also to be demonstrated that, with the selected limiting span length, large amplitude crossflow motion will not occur. A free span is to be corrected by appropriate means whenever the strength and fatigue criteria specified in Chap. 8 are not met. A span rectification plan is to be established for correcting pipeline spans. Spans are to be supported at the midpoint, where possible, or along an extended portion of the span. A detailed design is to be prepared for span supports that may be used on a range of span clearances. The support is to ensure the stability of the pipeline and to incorporate sufficient flexibility in its design to enable it to be modified in the field to suit the requirements of each span. Alternatively, techniques including sandbags, grout bags, mattresses, trenching, rock dump or combinations of different methods may be used for span correction.

9.5   Upheaval and Lateral Buckling

Pressure and temperature from flow contents may cause expansion in the pipe length and the pipe may buckle to a new equilibrium position. Examples of expansion include vertical downwards in a free span, up-lift on a free span shoulder and upheaval buckling for buried lines, and horizontal snaking and/or lateral buckling on the seabed. The following may accelerate global buckling,

- Uneven seabed,
- Local reduction in friction resistance, and
- Fishing gear impact, pull-over and hooking loads.

Pipes under internal/external pressure and seabed friction may be modelled as compressive beam columns subject to an "effective force". The seabed friction factors are to be properly modelled in the analysis.

9.5.1   Upheaval Buckling

In the case of an upheaval buckling resulting in limited plastic bending strains, pipelines may continue operating, provided that the operating parameters are kept within a range that prevents the accumulation of low-cycle high strain fatigue damage in the buckled section and that local buckling will not occur. When applicable, local buckling, out-of-roundness and fatigue damage analyses are to be conducted and submitted for approval by Class.

9.5.2   Lateral Buckling

Allowing lateral buckling may be an effective way of designing high pressure/high temperature pipelines, provided that the pipeline is not at risk of local buckling or at risk of colliding with another pipeline or installation. In such situations, cyclic and dynamic behaviour of the pipes is to be documented through simulation of in-situ behaviour. When applicable, local buckling, out-of-roundness and fatigue damage analyses are to be conducted and submitted for approval by Class.

9.6 Design for Impact Loads

For unburied pipelines and buried pipelines where scour conditions could remove the overburden, accidental impact loading from such activities as anchoring, fishing operations, etc., is to be considered. The energy absorption properties of the pipeline are to be determined and correlated with the probable accidental loading. Pipeline design is to consider impact loads, such as:

- Fishing gear impact, pull-over and hooking loads (when applicable),
- Dropped objects, and
- Anchoring.

9.6.1 Fishing Gear Loads

When it is necessary to design for fishing gear loads, the weight, sizing (length and breadth) of the fishing gear and trawl velocity for the route where pipelines are to be installed are to be investigated to form a design basis. Due consideration is to be given to the future developments or changes in equipment within the lifetime of the pipelines.

9.6.2 Dropped Object(s)

Design for dropped objects is to be conducted accounting for falling frequency, weight, and velocity of the dropped objects. The methods of analysis and acceptance criteria may be similar to those defined for fishing gear impact.

9.6.3 Anchoring

Design for anchoring is to be conducted, accounting for falling frequency, weight, velocity of the dropped objects and different scenarios causing anchor drop/drag on the unburied pipelines. The methods of analysis and acceptance criteria may be similar to those defined for fishing gear hooking loads.

Routing, Installation, Construction and Testing

10

10.1 General

When selecting the pipeline route, consideration is to be given to the safety of all parties, environmental protection, and the likelihood of damage to the pipe or other facilities. Any future activities in the immediate region of the pipeline are to be accounted for when selecting the route. In selecting a satisfactory route for an offshore pipeline, a field hazards survey may be performed to identify potential hazards, such as sunken vessels, pilings, wells, geologic and manmade structures, and other pipelines. Appropriate regulations are to be applied for minimum requirements for conducting hazard surveys. The selection of the route is to account for applicable installation methods and is to minimise the resulting in-place stresses.

The pipelines will run from a pipeline end manifold base or Pipeline End Manifold (PLEM) structure beneath the Floating Production Installation (FPI) to an unspecified destination. The pipeline corridor is to be selected based on hydrographic survey data. Hydrographic survey is to identify seabed topography bathymetry, seabed features, wrecks and debris. Geotechnical and geophysical survey is to be performed to determine surface and subsurface strata characteristics. The soil bearing capacity of surface and subsurface soil layers is to be assessed. Significant soil weakness or instability may necessitate routing realignment or design measures to enhance pipeline stability. Export line route selection is to endeavour to minimise deviation since additional length may escalate cost. A seismic study, if applicable, is to highlight the geohazard potential of unstable features.

In the route selection process, it is to be recognised that factors such as safety and functional integrity are to take precedence. A more direct and cost-effective route may be selected, even though a shorter routing may place the pipeline at a higher level of risk from external factors. In such a case, appropriate measures are to be taken within the design of the system to mitigate possible adverse effects. All physical constraints,

A. A. Olsen, *Subsea Pipeline Systems*, Synthesis Lectures on Ocean Systems Engineering, https://doi.org/10.1007/978-3-031-74790-8_10

both natural and man-made, should be identified along the pipeline corridor. Current and planned development, both along the pipeline corridor and in the immediate vicinity, are to be identified. A distance of 100 m is to be observed between the pipeline and an existing subsea facility. A separation distance of 30 m is to be maintained in between any existing or planned pipelines unless the pipeline is to be crossed or the pipeline approaches the platform, in which case, the spacing may be gradually reduced to that of the riser facing of the platform. Special route surveys may be required at landfalls to determine:

• Environmental conditions caused by adjacent coastal features,
• Location of the landfall to facilitate installation, and
• Location to minimise environmental impact.

10.1.1 Route Survey

The route survey covers survey for design purposes, survey for pre-installation and as-laid survey. Issues related to the seabed intervention are discussed in Chap. 9.

10.1.1.1 Route Survey for Design

A detailed route survey is to be performed for the planned pipeline route to provide sufficient data for the design. The width of the survey corridor is to be wide enough to cover the installation tolerance. Detailed and more accurate surveys are required, especially for uneven seabed topography, obstructions, near installations or templates, existing pipeline or cable crossings, subsurface conditions, large boulders, etc.

Seabed bathymetry and soil data properties are to be investigated and provided based on the route survey results. Seabed properties, including different soil layers, are to be included in the route survey maps. As a minimum, the soil stiffness and seabed soil friction coefficients in both axial and lateral directions are to be provided.

10.1.1.2 Route Survey for Pre-installation

Prior to installation, the route is to be surveyed for the following cases:

• New installations along the route,
• Changes of installation (e.g., templates) location,
• Change of seabed conditions due to heavy marine activities, and
• New requirements due to seabed intervention design or installation engineering.

10.1.1.3 As-Laid Survey

After installation, an as-laid survey is to be conducted. As a minimum, the following is to be included in the as-laid survey:

- Position (coordinates) and water depth profile for the pipeline systems,
- Coordinates identifying starting and ending point,
- Identification and quantification of any spans, crossings, structures, and large obstructions, and
- Reporting of any damages to the pipeline systems, including pipes, cathodic protection system, weight coating, supports, appurtenances, in-line structures, etc.

If necessary, an integrity assessment of the pipeline system based on the as-laid survey is to be performed.

10.1.1.4 As-Built Survey

After the final testing of the pipeline system, an as-built survey is to be conducted. This survey may be limited to key locations defined during the as-laid survey, such as identified spans, crossings, subsea structures, and areas with special features. As a minimum, the following is to be included in the as-built survey:

- Positions (coordinates) and water depth profile, including tangential points, out-of-straightness measures, etc.,
- Quantification of spans, crossings, trench, burial, scour, erosion, and
- Reporting of any damages to the pipeline systems, including pipes, cathodic protection system, weight coating, supports, appurtenances, in-line structures, etc.

10.1.2 In-Field Pipelines

The in-field area is subject to detailed geotechnical and geophysical survey to determine soil and sub- structural characteristics, local surface irregularities and obstructions. In-field pipeline routing is to consider vessel mooring patterns and temporary anchor locations of Mobile Offshore Drilling Units (MODUs). The potential of interference between mooring lines and risers/pipelines under extreme vessel motion is to be assessed. In developing pipeline layouts, the selected minimum route curvature radius is to be verified such that it is sufficient to maintain the pipeline from moving when the lay barge executes route turn. Bending capacity of the pipe, as described in Chap. 8, is also to take account of the selection of minimum route curvature radius. The soil lateral friction can maintain the curvature of radius R during installation, if:

$$R \geq \frac{T}{\mu \cdot W_s}$$

where:

R route curvature radius along the route

T   on-bottom lay tension
μ   soil lateral friction coefficient
W_s submerged weight of the pipe

Pipeline installation sequence and construction method are to be reviewed to determine the influence on pipeline layout in the field area. The feasibility of all proposed pipeline layouts and installation/tie-in methods is to be confirmed.

10.2   Installation Analysis

An analysis of the pipe-laying operation is to be performed, taking account of the geo-metrical restraints of the anticipated laying method and lay vessel, as well as the most unfavourable environmental condition under which laying will proceed. The analysis is to include conditions of starting and terminating the operation, normal laying, abandonment and retrieval operation, and pipeline burial. The analysis is to ensure that excessive strain, fracture, local buckling, or damage to coatings will not occur under the conditions anticipated during the pipe-laying operation. Strength analysis is to be performed for the pipeline during laying and burial operations. The strength analysis is to account for the combined action of the applied tension, external pressure, and bending and dynamic stresses due to laying motions, when applicable.

10.2.1 S-Lay Installation

For S-lay installation, the pipe is laid from a near-horizontal position using a combination of horizontal tensioner and a stinger controlling the curvature at over-bend. The lay vessel can be a ship, barge, or a semi-submersible vessel. The required lay tension is to be determined based on the water depth, the submerged weight of the pipeline, the allowable radius of curvature at over-bend, departure angle and the allowable curvature at the sag-bend. The stinger limitations for minimum and maximum radius of curvature and the pipeline departure angle are to be satisfied. Strain concentrations due to increased stiffness of in-line valves are to be accounted for. The in-line valve is to be designed for strength and leakage protection to ensure the integrity of the in-line valve after installation.

Due to local increased stiffness by external coatings and buckle arrestors, for example, strain in girth welds may be higher than in the rest of the pipe, and strain concentration factors are to be calculated based on strain level and coating thickness or wall-thickness of buckle arrestors. Installation procedures are to safeguard the pipe with coatings, protection system, valves and other features that may be attached. A criterion for handling the pipe during installation is to consider the installation technique, minimum pipe-bending radii, differential pressure, and pipe tension.

10.2.2 J-Lay Installation

For J-lay (near-vertical pipe-lay), the pipe is laid from an elevated tower on a lay vessel using longitudinal tensioner. In this way, over-bend at the sea surface is avoided. In general, J-lay follows the same procedure as S-lay.

10.2.3 Reel Lay Installation

For reel lay, the pipe is spooled onto a large radius reel aboard a reel lay vessel. The reeloff at location will normally occur under tension and involve pipe straightening through reverse bending on the lay vessel. The straightener is to be qualified to ensure that the specified straightness is achieved. Anodes are, in general, to be installed after the pipe has passed through the straightener and tensioner. Filler metals are to be selected to ensure that their properties after deformation and aging match those of the base material. Fracture mechanics assessment may be conducted to assess ductile crack growth and potential unstable fracture during laying and in service. The allowable maximum size of weld defects may be determined based on fracture mechanics and plastic collapse analysis.

10.2.4 Installation by Towing

The pipe is transported from a remote assembly location to the installation site by towing either on the water surface, at a controlled depth below the surface or on the sea bottom. The submerged weight of the towed pipeline (e.g., bundles) is to be designed to maintain control during tow. The bundles may be designed to have sufficient buoyancy by encasing the bundled pipelines, control lines and umbilical inside a carrier pipe. Ballast chains may be attached to the carrier pipe at regular intervals along the pipeline length to overcome the buoyancy and provide the desired submerged weight. In the case of bottom tow installation, the route is to be surveyed carefully, and the pipe must have an abrasion-resistant coating that can stand up to dragging across the seabed.

10.3   Construction

Pipelines are to be constructed in accordance with written specifications that are consistent with this guide. The lay methods described in this chapter and other construction techniques are acceptable, provided the pipeline meets all the criteria defined in this guide. Plans and specifications are to be prepared to describe alignment of the pipeline, its design water depth, and trenching depth and other parameters.

10.3.1 Construction Procedures

The installation system is to be designed, implemented, and monitored to ensure the integrity of the pipeline system. A written construction procedure is to be prepared, including the following basic installation variables:

- Water depth during normal lay operations and contingency situations,
- Pipe tension,
- Pipe departure angle,
- Retrieval, and
- Termination activities.

The construction procedure is to reflect the allowable limits of normal installation operations and contingency situation.

10.3.2 Buckle Detection

Detections of dents, excessive ovality or buckles in the pipeline are to be performed during pipe laying. Whenever possible, the detection is to be accomplished by passing a buckle detector through the pipe section. Alternative methods capable of detecting changes in pipe diameter may be used upon agreement with Class.

10.3.3 Weld Repair

For weld repair carried out at weld repair stations, weld repair analysis is to be performed to ensure that the length and depth of cut combination does not produce combined stresses that exceed the allowable stress during pipe laying.

10.3.4 Trenching

In the event on-bottom stability of the pipeline cannot be achieved with a means of shielding the pipeline from the effects of current, the impact on the pipeline of trenching is to be determined. Pipeline trenching may require lifting of the pipeline to allow clearance for plow shares to excavate a trench. The pipeline minimum allowable curvature for trenching is to be determined and the pipeline vertical curvature of the as-installed pipeline is to be as large as possible to control the pipeline stresses within allowable limits during operational phase. If required, the design is to include stress analysis of this configuration to ensure that stresses remain within allowable limits. The standard depth of trenching for pipelines is the depth that will provide 0.9 m (3 feet) of elevation differential between the top of the pipe and the average sea bottom. The hazards are to be evaluated to determine the total depth of trenching in those situations where additional protection is necessary or mandated. A post-trenching survey is to be conducted to determine if the required depth has been achieved.

10.3.5 Gravel Dumping

Gravel dumping is to be controlled such that the required gravel is dumped over and under the pipeline and subsea structures and over the adjacent seabed. During the gravel dumping operations, inspection is to be carried out to determine the performance of the dumping. Measures are to be taken to avoid damaging the pipeline and coating during the dumping process. Upon completion of the gravel dumping, a survey is to be conducted to confirm the compliance with the specified requirements.

10.3.6 Pipeline Cover

Pipeline cover is normally installed where more protection is required. Pipeline and coating are to be protected from damage in areas where backfill is specified, or where pipeline-padding material is specified.

10.3.7 Pipeline Crossings

In deepwater locations, pipeline crossing is not normally required. In-field pipeline layouts are to ensure that crossing is avoided where feasible and export lines are to be deviated such that the potential for interference is avoided. Where crossings cannot be avoided, a crossing design is to be developed that accounts for the accuracy with which pipelines can be installed in extreme depths. Vertical separation at crossings between new and existing pipelines is to be at least one (1) foot, as required in ASME 31.4. The pipeline crossing profile is to be checked for in-place stresses for hydrotest, operational, and environmental loads. The stability of supports is to be checked for sliding and overturning moments. Crossing design is also to account for soil bearing strength and the requirement for additional support to minimise or avoid the creation of an unsupported span.

10.3.8 Protection of Valves and Manifolds

Valves, manifolds, and other subsea structures installed on an offshore pipeline are to be protected from fishing gear and anchor lines. Protective measures are to be applied to prevent damage to the valves and manifolds. Such measures are not to obstruct trawling or other offshore operations.

10.3.9 Shore Pull

Shore pull is a process in which a pipe string is pulled either from a vessel to shore or vice versa. Installation procedures are to be prepared, including installation of pulling head, tension control, twisting control and other applicable items. Cables and pulling heads are to be dimensioned for the loads to be applied, accounting for overloading, friction, and dynamic effects. Winches are to have adequate pulling force and are to be equipped with wire tension and length indicators. Buoyancy aids are to be used if required to keep pulling tension within acceptable limits. Buoyed pipeline sections' lateral stability is to be analysed for installation phase.

10.3.10 Tie-In

Tie-in procedures are to be prepared for the lifting of the pipeline section, control of configuration and alignment, as well as mechanical connector installation. Alignment and position of the tie-in ends are to be within specified tolerances prior to the tie-in operation.

10.3.11 Shore Approaches

The choice of shore crossing location, design and construction technique is to take account of the following factors:

(1) Changing nature of the shorelines,
(2) Environmental importance of shorelines,
(3) Complexity of sea/land interface, and
(4) Existing pipelines, cables, and outfalls in the area.

Adequate site investigation and knowledge of environmental conditions are to be obtained in planning shore approaches. Marine survey is to be carried out to determine the shore profile, the ocean and tidal currents and the seabed bathymetry, which determines local wave refraction. Geotechnical site investigation is to be carried out to determine the geotechnical description and strength properties of the seabed material.

Special Considerations for Pipe-In-Pipe Design

11

11.1 General

This chapter defines the design criteria that are specifically applied to pipe-in-pipe systems. Relevant failure modes for pipelines, described in Chap. 8, are to be considered in the design of pipe-in-pipe systems.

11.2 Design Criteria

11.2.1 Strength Criteria

The design of pipe-in-pipe systems is, in general, to follow the strength criteria given in Chap. 8. For high temperature/pressure systems, an equivalent strain criterion may be applied as:

$$\varepsilon_p = \sqrt{\frac{2\left(\varepsilon_{p\ell}^2 + \varepsilon_{ph}^2 + \varepsilon_{pr}^2\right)}{3}} =$$

where:

ε_p = longitudinal plastic strain.
ε_{ph} = plastic hoop strain.
ε_{pr} = radial plastic strain.

The maximum allowable accumulation of plastic strain is to be based on refined fracture calculations and is to be submitted for approval by Class. The inner pipe burst capacity of

© The Author(s), under exclusive license to Springer Nature Switzerland AG 2025
A. A. Olsen, *Subsea Pipeline Systems*, Synthesis Lectures on Ocean Systems
Engineering, https://doi.org/10.1007/978-3-031-74790-8_11

the pipe-in-pipe system is determined based on the internal pressure, and local buckling capacity is evaluated based on the outer pipe subjected to the full external pressure.

11.2.2 Global Buckling

In terms of structural behaviour, a pipe-in-pipe system may be categorised as being either compliant or non-compliant, depending on the method of load transfer between the inner and outer pipes. In compliant systems, the load transfer between the inner and outer pipes is continuous along the length of the pipeline, and no relative displacement occurs between the pipes, whereas in non-compliant systems, force transfer occurs at discrete locations. Due to effective axial force and the presence of out-of-straightness (vertically and horizontally) in the seabed profile, pipe-in-pipe systems are subjected to global buckling, namely, upheaval buckling and lateral buckling. In contrast to upheaval buckling, lateral buckling may be accepted if it does not result in unacceptable stresses and strains. In case global buckling may occur, a detailed finite element analysis is recommended to further investigate the buckling behaviour.

11.2.3 Pipe-In-Pipe Appurtenances Design

Design of pipe-in-pipe appurtenances is to be based on their functional and installation requirements including field joint design, bulkhead design, spacers, water stops/intermediate bulkheads to prevent water intrusion at the field joints, and insulation.

11.3   Special Considerations for Pipeline Bundle Design

This section defines the design criteria that are specifically applied to pipeline bundle systems. Relevant failure modes described in Chap. 8, are to be considered in the design of pipeline bundle system.

11.3.1 Functional Requirements

Pipeline bundle systems may be used in applications when the same end points have to be connected by more than one pipeline or where pipe insulation is required to avoid hydrate and wax formation. A bundle can be open, with the individual pipes and cables strapped together, or closed, with all of them contained in an outer carrier pipe. The following functional requirements are to be considered.

11.3.2 Design Temperature/Pressure

Design temperature and pressure are to be based on hydraulic analyses. Significant temperature drops along the bundle system are to be avoided.

11.3.3 Pigging Requirements

If the bundle system is designed for pigging, the geometric requirement is to be fulfilled. The minimum bend radius is usually five (5) times the nominal internal diameter of the pipe to be pigged.

11.3.4 Settlement/Embedment

Settlement/embedment of the pipeline bundle system is to be accounted for, where appropriate.

11.3.5 System Thermal Requirements

The insulation value is to be determined based on thermal conduction, convection and hydraulic analysis considering:

- Maximum/minimum operating temperature,
- Cool down time,
- Heat-up time,
- Heat-up system volume,
- Bundle configuration and the relative positions of the components,
- Bundle length,
- Heating medium temperature,
- Heating medium flow rate, and
- Properties of the fluid contained inside the pipelines.

11.4   Design Criteria for Bundle Systems

The design of bundle systems is to ensure that the system satisfies the functional requirements, and adequate structural integrity is maintained against all the failure modes. Design of the carrier pipelines, heat-up lines and service line is to be in accordance with Chap. 8. The design criteria are to be applicable for installation, system pressure test and operation.

11.4.1 Carrier Pipe Design

Carrier pipe stresses are to be checked during the carrier selection and are to be reviewed in light of the stresses determined from the expansion analysis. The weight of all the bundle component parts is to be determined. The displacement of external anodes, clamps and valves is to be accounted for by using a submerged weight for these items in the weight calculations.

11.4.2 Wall Thickness Design Criteria

Pipelines are to be sized according to processing data. The wall thickness of the pipelines depends on the internal pressure containment and the pressure in the annulus. For high temperature applications, thermal loading is to be considered in pipeline sizing and selection of material properties that will apply at elevated temperatures. The wall thickness design of pipes within the bundle system is to account for the following.

11.4.2.1 Hoop Stress
The hoop stress criterion in Chap. 8 is applicable for bundle systems.

11.4.2.2 Collapse Due to External Hydrostatic Pressure
The wall thickness of the carrier pipe is to be designed to avoid pipe collapse due to external hydrostatic pressure.

11.4.2.3 Local Buckling
The pipelines, carrier and sleeve pipe are to be designed to withstand local buckling due to the most unfavourable combination of external pressure, axial force, and bending.

11.4.2.4 Installation Stress
The wall thickness is to be adequate to withstand both static and dynamic loads imposed by installation operations.

11.4.2.5 System Pressure Test and Operational Stresses
The wall thickness is to be adequate to ensure the integrity of carrier and sleeve pipe under the action of all combinations of functional and environmental loads experienced during system pressure test and operation.

11.4.3 On-Bottom Stability Design

The bundle system is to be stable on the seabed under all environmental conditions encountered during installation, testing and throughout the design life. The bundle is to be designed with sufficient submerged weight to maintain its installed position, or to limit movement such that the integrity of the bundle system is not adversely affected. The bundle may be considered stable when actual submerged weight is greater than the minimum required multiplied a factor of 1:1. On-bottom stability is to be verified in accordance with Chap. 9.

11.4.4 Free Spans Design

Maximum allowable span length for the bundle system is to be calculated for both static and dynamic loading conditions. Analyses are to consider all phases of the bundle design life, including installation, testing and operation.

11.4.5 Bundle Expansion Design

The design is to account expansion and/or contraction of the bundle as a result of pressure and/or temperature variation. The bundle expansion analysis includes determination of external and internal forces acting on the system, calculation of axial strain of the system and integration of the axial strain of unanchored bundle to determine the expansion. Design pressure and the maximum design temperature are to be used in bundle expansion analysis. The presence of the sleeve pipe is to be accounted for. Bundle expansion movement is to be accommodated by the tie-in spools.

11.4.6 Bundle Protection Design

The bundle system is to be designed against impact loads in areas where fishing activities are frequent. The pipelines are to be protected against dropped objects around the installations (refer to Chap. 9). Carrier pipe and bundle towheads are to offer sufficient protection against the dropped objects.

11.4.7 Corrosion Protection Design

Cathodic protection design is to be performed according to relevant codes, e.g., NACE RP0176. Cathodic protection together with an appropriate protective coating system is to be considered for protection of the external steel surfaces from the effects of corrosion. The sleeve pipe is to be protected from corrosion using chemical inhibitors within the carrier annulus fluid. Pipelines within the sleeve pipe are to be maintained in a dry environment, and a cathodic protection system will therefore not be required.

11.4.8 Bulkheads and Towhead Structure Design

The bulkheads will form an integral part of the towhead assemblies. The towhead structure is to remain stable during all temporary and operational phases. Stability is to be addressed with respect to sliding and overturning with combinations of dead weight, maximum environmental and accidental loads applied. The design of towhead structure is to be in accordance with relevant structural design code.

11.4.9 Bundle Appurtenances Design

Design of bundle appurtenances is to be based on their functional requirements, including spacers, filling, flood, and vent valves, transponder supports, chain attachment straps, and tie-in spools.

Testing, Drying and Commissioning

12

12.1   General

This chapter describes the minimum functional requirements to be met during hydrostatic test and commissioning of a pipeline system composed generically of a subsea pipeline and all connected interface components. The purpose of the commissioning is to prepare the pipeline system for the acceptance of production hydrocarbons, lift gas and injection water. For an oil pipeline system, commissioning is mainly dewatering. For the gas pipelines, drying is normally required before introduction of the hydrocarbons into the pipeline system. Additional project-specific commissioning procedures are also likely to need to be developed. Accordingly, the minimum functional requirements include the following:

- Cleaning, gauging and line filling,
- Hydrostatic pressure test,
- Leak test (if required),
- Post-testing and rectification requirements,
- Pigging requirements,
- Dewatering, and
- Drying requirements.

Hydrostatic pressure test is to be performed on the completed system and on all components not tested with the pipeline system or components requiring a higher test pressure than the remainder of the pipeline. If leaks occur during tests, the leaking pipeline section or component is to be repaired or replaced and retested in accordance with this guide. Detailed procedures for the hydrostatic testing and commissioning of the pipeline system are to be developed. The procedures are to be such that they do not jeopardise the pipeline

© The Author(s), under exclusive license to Springer Nature Switzerland AG 2025
A. A. Olsen, *Subsea Pipeline Systems*, Synthesis Lectures on Ocean Systems
Engineering, https://doi.org/10.1007/978-3-031-74790-8_12

system's fitness for purpose, e.g., by overstressing the whole pipeline system or parts of it. In addition, the procedures are not to impose additional requirements over and above specified operational conditions on mechanical design of the pipeline.

12.2   Testing of Short Sections of Pipe and Fabricated Components

Short sections of pipe and fabricated components such as scraper traps and manifolds may be tested separately from the pipeline. Where separate tests are used, these components are to be tested to pressures equal to or greater than those used to test the pipeline system.

12.2.1 Testing After New Construction

12.2.1.1   Testing of Systems or Parts of Systems
Pipelines designed according to this guide are to be system pressure-tested after completion of trenching, gravel dumping, covering, and crossing. The test is to be performed after installation and before operation. It is to be ensured that excessive pressure is not applied to valves, fittings, and other equipment. The valve position and any differential pressure across the valve seat are to be specifically defined in the test procedures.

12.2.1.2   Testing of Tie-Ins
Given it is sometimes necessary to divide a pipeline system into test sections and install weld caps, connecting piping and other test appurtenances, it is not required to test tie-in welds. However, tie-in welds that have not been subjected to a pressure test are to be radiographically inspected or subjected to other accepted non-destructive methods. After weld inspection, field joints are to be coated and inspected. Mechanical coupling devices used for tie-in are to be installed and tested in accordance with the manufacturer's recommendations.

12.3   Cleaning, Gauging and Line Filling

Prior to pressure testing, the pipeline system is to be cleaned and gauged to ensure removal of construction debris and loose scales and to check that the pipeline system is free of deformations and/or obstructions. Prior to using water for the cleaning operation, all pipeline spans must be inspected to ensure that the unsupported spans do not exceed the allowable span value for hydrostatic test conditions. Filling the pipeline for hydrostatic testing normally takes place as part of the gauging operation. The water used is to be

fresh water or filtered seawater, depending on the project-specific requirements. It is recommended that the air content in the system during the pressure test does not exceed
0.2% of the volume of the system being tested.

12.3.1 Cleaning

The first pig driven through the cleaning section is to be of a bi-directional type. The
position of the pig should be monitored. All debris received with the pig is to be disposed
of in an authorised manner.

12.3.2 Gauging

The test section is to be gauged after cleaning. A bi-directional pig is to be fitted with
one or two aluminium gauging plates. The diameter of the gauging pig is recommended
to be 95% of the inner drift diameter of the pipe. A record is to be taken of the condition
of each gauging plate before and after use. Any damage noted by the gauging pig is to
be located and repaired.

12.3.3 Line Filling

Necessary measures are to be taken to remove air from the line during filling. The filling
speed should be specified prior to the filling operation. In most cases, the filling pig speed
is to be approximately 2 ft/s and is not to exceed 6 ft/s. To assist in controlling the line
filling and water treating, the following measurements and records are to be taken:

- Inlet flowrate,
- Inlet pressure,
- Inlet temperature,
- Chemical injection rate (if carried out),
- Dye injection rate (if carried out), and
- Precautions are to be taken during freezing conditions.

12.3.4 Temperature Stabilisation

The temperature of the line-fill water should be stable before testing commences. The
calculation of the temperature stabilisation period is to be detailed in the test procedure.

Pressures and temperatures, including ambient, are to be recorded regularly during the stabilisation period.

12.3.5 Pressurisation

The asset owner is required to provide the Class Surveyor with a chart that shows a graph of pressure versus added volume (P/V plot) using measurement of volume added either by pump strokes or flow meter and instrument reading of pressure gauge and a dead weight tester. The rate of pressurisation should be constant and not exceeding 1 bar per minute until a pressure of 35 bar or 50% of the test pressure, whichever is lesser, has been reached. During this period, volume and pressure readings should be recorded at regular intervals.

When the pressure of 35 bar or 50% of the test pressure, whichever is lesser, has been reached, the air content in the test line is to be determined. When the air content is within the maximum allowable limit of 0.2% of the test section volume, the pressurisation should continue. The pressures and added volumes should be continuously plotted until the specified test pressure has been reached.

12.4   Hydrostatic Pressure Testing

After installation and before operation, all parts of an offshore pipeline designed according to this guide are to be subjected to hydrostatic test. API RP 1110 and API RP 1111 may be used as guidance on the hydrostatic test. The system pressure test is not to result in hoop stress and combined stress exceeding the capacity given in Chap. 8. Precautions are to be taken to ensure safety of personnel during the entire test procedure. The test medium is to be fresh water or seawater unless freezing may happen. Corrosion inhibitor and biocide additives are to be added to the test medium in case the water is to remain in the pipeline for an extended period. Effects of temperature changes are to be accounted for when interpretations are made of recorded test pressures. Plans for the disposal of test medium together with discharge permits are to be acceptable to the local authorities. The purpose of the hydrostatic pressure test of the pipeline system is to ensure its mechanical strength after completion of construction and to verify that the system is leak free.

The pressure and the duration of the test are to be in accordance with the requirements specified in this chapter. For the purpose of this functional requirement, it is assumed that the hydrostatic pressure test is to fulfil the requirements of ASME B31.8 (gas pipeline system) and ASME B31.4 (liquid pipeline system) as a minimum. A detailed hydrostatic pressure test procedure is to be developed for the purpose of testing the pipeline system.

12.4.1 Assumptions and Sequence of Operations

The hydrostatic test of the pipeline system is performed after completion of all installation and construction work and before operation. The extent of the test is to be based on actual project configuration and it is to include the entire pipeline and installed interface components as a minimum. Depending on the offshore installation scenarios, some components of the pipeline system, such as risers, may be subject to a local hydrostatic leak test prior to the hydrostatic test of the whole pipeline system. It is recommended that the connected components and pipeline system be tested as a unit. The sequence of pressure testing operations of the pipeline system is to be as follows:

(1) Filling,
(2) Cleaning and gauging,
(3) Hydrotesting, including temperature stabilisation, pressurisation, air contents check and hydrostatic test/holding period,
(4) Post-testing, including depressurisation and documentation,
(5) Rectification activities (if required), including leak location during test, dewatering for rectification and rectification of defects,
(6) Final/repeat hydrotesting, and
(7) Testing, certificates witness signature.

12.4.2 Hydrotest Pressure and Duration

Hydrostatic pressures both internally and externally are to be accounted for. Guidance on the work procedure can be found in API RP 1111 and API RP1110. The minimum duration of the hydrostatic pressure test is to be a strength test at the test pressure for at least eight (8) continuous hours. The assembly testing is to be comprised of a four-hour strength test followed by visual examination at the leak tightness test pressure. A leak tightness test may be combined with strength test or be commenced immediately after strength test has been completed satisfactorily.

The test pressure at any point of the test section is not to be less than 1.25 times of the maximum allowed operating pressure (MAOP) and at least be equal to the test pressure required in the ANSI/ASME B31.4 or B31.8, or to the pressure creating a hoop stress of 90% SMYS of the line pipe material, based on the minimum wall thickness, whichever is higher. During the hydrostatic pressure test, the combined stress is not to exceed 100% SMYS of line pipe material based on minimum wall thickness.

The margin between the hoop stress of 90% SMYS and the combined stress of 100% SMYS allows for elevation differences in the test section and/or longitudinal stresses, e.g., due to bending. However, the elevation differences in each test section are to be limited to a value corresponding to 5% of SMYS of the line pipe material or to 50 m or as specified in the scope of work. It is to be confirmed that the calculated test pressure does not exceed the design pressure of the fittings specified for the pipeline.

The combined stress for the hydrostatic pressure test condition is to be calculated in accordance with the guidance outlined in this book. The calculation is to include major residual stresses from construction, e.g., from towing or lay barge operations and longitudinal stresses due to axial and bending loads, e.g., at unsupported pipeline spans. The combined stress during the hydrostatic pressure test is to be limited to 100% of SMYS based on the minimum wall thickness. If the calculated combined stress is higher than 100% of SMYS, special measures are to be taken to reduce the longitudinal stresses in the test section. The pressure is to be maintained during the strength test at test pressure ±1 bar by bleeding or adding water as required. The volumes of water added or removed are to be measured and recorded. The test section temperature and the ambient temperature against time plot created for the stabilisation period is to be maintained.

12.4.3 Leak Tightness Test

The leak tightness test may be commenced immediately after the strength test has been completed satisfactorily or combined with strength test. No water is to be added or removed during the tightness test. The test is intended to demonstrate that there is no leak in the pipeline. To allow for pressure variations caused by temperature fluctuations during the test duration, the leak test pressure is to be set to a level of 80% of the hydrostatic test pressure if they are carried out separately. If it can be ensured that the pressure variations due to temperature fluctuations are within the specified limits, a combined strength/leak tightness test at the strength test pressure, without water addition or removal, is to be carried out.

12.4.4 Acceptance Criteria

The pressure variation is not to exceed ±0.2% of the test pressure during the test period. Due to temperature changes, a pressure variation of up to ±0.4% of the test pressure may be acceptable. The pressure test is only acceptable when both the above pressure variation criteria are met, and no leakage is observed. To determine whether any pressure variation is a result of temperature changes or whether a leak is present, the pressure and or temperature changes are to be calculated from an appropriate pressure and temperature equation formula for an unrestrained test section.

12.4.5 Pipeline End Manifold (PLEM) Hydrotest Considerations

The PLEM's pipework will be subjected to a separate hydrostatic test prior to the off-shore leak tightness test. PLEM can be treated as an assembly/fabricated item. During the offshore test, care is to be exercised to ensure that excessive pressure is not applied to valves, fittings, and other components.

12.5   Post-test and Rectification Requirements

After the satisfactory completion of the hydrostatic pressure test, the pipeline system is to be depressurised. The depressurisation should be performed at a steady and controlled rate. The manufacturer of the connected components defines the maximum depressurisation rate. If a leak is suspected, the pressure is to be reduced to less than 80% of the test pressure before carrying out a visual examination. If it is not possible to locate the suspected leak by visual examination, a method is to be used such that the locating of leaks can be done at test pressure without endangering the personnel carrying out the work. When the leak has been found, the test section is to be repaired.

12.5.1 Dewatering for Rectification

To rectify any defects, it may be necessary to partially or completely dewater the test section. Prior to dewatering, it is to be confirmed that all block valves, if installed, are in the fully open position. The disposal of line-fill water is to be done in accordance with accepted procedures. The test section is not to be left in the partially or completely dewatered condition longer than one week without any further internal corrosion protection. Depending on the post-dewatering period and the line-fill water quality, it may be required to purge nitrogen or swab the test section with fresh water and/or inhibition slugs to avoid internal corrosion in the test section.

12.5.2 Rectification of Damage

Damage and/or defects detected during cleaning, filling, gauging and pressure test operations are to be located and repaired or replaced in accordance with the appropriate pipeline construction specification. The proposed repair and/or replacement procedure is to be verified and approved when the actual site conditions are known. Following the completion of repairs, the activity, which has been interrupted for the repair works, should be repeated until the failed activity has been successfully completed.

12.6   Pigging Requirements

Pigging operations are to comply with the activity-specific requirements specified in the offshore installation procedures.

12.6.1 Pig Selection

Cleaning pigs with steel brushes or brush-coated foam pigs are to be used in uncoated carbon steel pipelines. Only polyurethane plates, disc pigs, spiral wound or criss-cross polyurethane coated foam pigs are to be used for cleaning of internally flow coated pipelines. Pigs are to be suitable to pass through pipelines with minimum bore and minimum bend radius, as specified in the scope of work. Foam pigs are to be oversized as required for their duty. Brush or abrasive coated foam pigs should replace bare foam pigs if inspection after pigging reveals pig deterioration to the extent that they may disintegrate during pigging and/or become inefficient for dewatering.

12.6.2 Planning Pigging Operations

Pipeline elevation, density of fluid in the pipeline and back-pressure at the receiving end are to be accounted for when calculating the required inlet flows and pressures for the driving medium. Prior to any pigging operations, it is to be verified that:

(1) All valves have been correctly positioned and are functioning,
(2) All main line valves are to be in the fully open position, and
(3) Discharge and/or storage facilities have been connected and are in good working order Pigs should be removed from the pig receiver immediately upon arrival and then inspected.

12.6.3 Pig Train Monitoring

During all pigging operations, the location of the pig trains should be predicted by calculations and measurement of the volume of the driving medium. Launching and arrival of pigs at pig traps are to be monitored by pig signallers fitted to the pig traps or by measuring signals from pig location devices fitted to the last pig of the pig train. When glycol or other chemicals are part of the pig train or used as the driving medium, the location of pig trains should be monitored during the pigging operations. Location monitoring should be done by measuring the signals from pig location devices or by predictions based on the volumes of the discharged fluids measured in the discharge pipe work. All pigging operations are to be recorded in a pig data register with the following data as a minimum:

(1) Number of the pig runs,
(2) Type(s) of pig,
(3) Volumes and pressures of the driving media,
(4) Time of launching and receiving,
(5) Condition of pigs before launching and upon arrival, and
(6) Volume(s) and nature(s) of material and substances arriving in front of the pig at the receiving end.

12.7 Dewatering

The pipeline system is to be dewatered with appropriate medium. The dewatering pigs are to be driven by nitrogen or by produced fluid, whenever practicable. The entrance of air/oxygen into the pipeline is to be avoided during the dewatering. The pipeline is not to be left dewatered with air for more than two (2) weeks unless otherwise permitted in the scope of work. Adequate size dump lines at the receiving end are to be provided, with respect to safety and legislation over control of pollution. The following are to be measured and recorded:

(1) Details of test water disposed from the receiving end,
(2) For each dewatering run the quantities, pressure, and flowrate records of the driving medium, and
(3) Dew point records of the air.

12.8 Drying of Gas Pipelines

The pipeline system is to be dried to a level that is specified for the intended service. One of the major factors determining the drying method is the dryness criterion. This dryness is based on the value that is required to avoid the formation of hydrates. The general requirements for drying are:

(1) Drying is to follow immediately after dewatering,
(2) The period between drying and commissioning is to be kept as short as possible,
(3) Only nitrogen is to be used as drying medium. Use of hydrocarbon gas as a drying medium should be avoided,

(4) If vacuum drying is going to be undertaken, it is important to verify if all the pipeline system components including PLEM valves and any risers, or other installed components are capable of vacuum service,

(5) Methanol/Glycol swabbing should be used when there may be a risk of a hydrate forming in the pipeline system, and

(6) Production gas drying can be used if the water content in the pipeline system following dewatering is sufficiently low to avoid the possibility of hydrate formation.

The records of critical parameters (dew point, temperature, pressure at inlet and outlet of the pipeline system, flowrate, total volume of drying medium, etc.) for each activity included in the drying process are to be provided for Class review.

Inspection, Maintenance and Repair 13

13.1 General

An inspection and monitoring philosophy are to be established, and this is to form the basis for the detailed inspection and monitoring programme. Inspection and monitoring are to be carried out to ensure safe and reliable operation of the pipeline system.

13.2 Inspection by Intelligent Pigging

The types and frequency of intelligent pig inspection are to be determined based on the asset Operator's inspection philosophy and the operational risks of the pipe system. The inherent limitations of each inspection tool are to be examined.

13.2.1 Metal Loss Inspection Techniques

Several techniques are applicable, for example:

- Magnetic flux leakage,
- Ultrasonic,
- High frequency eddy current, and
- Remote field eddy current.

13.2.2 Intelligent Pigs for Purposes Other Than Metal Loss Detection

Pipe inspection by intelligent pigging can be categorised into the following five groups of inspection capability:

- Crack detection,
- Calipering,
- Route surveying,
- Free-span detection, and
- Leak detection,

13.3   Monitoring and Control

Control systems such as these listed below are to be provided to ensure operational safety.

13.3.1 Emergency Shutdown

A means of shutting down the pipe system is to be provided at each of its initial and terminal points. The response time of an emergency shut down valve is to be appropriate to the fluid in the pipe (type and volume) and the operating conditions.

13.3.2 Pressure Protection

The pipe system is to be operated in a way that ensures the operating pressure is not exceeded. Primary overpressure protection devices which shut-in the production facilities (wells, pumps, compressors, etc.) are in no case to exceed the maximum allowable operating pressure. Secondary overpressure protection may be set above the maximum allowable operating pressure but is not to exceed 90% of System Test Pressure. Such primary and secondary protection will protect the pipeline and allow for the orderly shut-in of the production facilities in case of an emergency or abnormal operating conditions. In some cases, other overpressure protection device settings for subsea well pipelines may be allowed since in the case of an emergency the well(s) will be shut-in at the host facility by the emergency shutdown system. Instrumentation is to be provided to register the pressure, temperature, and rate of flow in the pipeline. Any variation outside of the allowable transients is to activate an alarm in the control centre.

13.3.3 Pressure, Temperature, and Flow Control

To ensure protection of the pipe system against over pressurisation and excessively high temperatures, automatic primary and secondary trips are to be installed. Details, including high/low pressure/temperature settings, are to be documented in the Operations Manual.

13.3.4 Relief Systems

Relief systems, such as relief valves, are typically required to ensure that the maximum pressure of the pipe system does not exceed a certain value. Relief valves are to be correctly sized, redundancy provided, and they are to discharge in a manner that will not cause fire, health risk or environmental pollution.

13.4   Maintenance

The principal function of maintenance is to ensure that the pipelines continue to fulfil their intended purpose in a safe and reliable way. Their functions and associated standards of performance are to be the basis for the maintenance objectives. Maintenance is to be carried out on all pipeline systems, including associated equipment (e.g., valves, actuators, pig traps, pig signallers and other attachments). Maintenance procedures and routines may be developed, accounting for previous equipment history and performance.

13.5   Pipeline Damage and Repair

In the event of pipe damage threatening the safe continuous transportation of hydrocarbons, inspection, reassessment, maintenance, and repair actions are to be promptly taken, as illustrated below:

- Identify possible cause of damage,
- Identify type of encountered damage,
- Define pipeline zone criticality and damage categorisation,
- Identify damage location and assessment techniques, and
- Outline repair techniques which may be applied to specific damage scenarios.

13.5.1 Categorisation of Damage

The causes of pipeline damage may be categorised as below.

13.5.2 Internal Damage

Internal corrosion damage occurring from the corrosivity of the transported product and flow conditions in combination with inadequate use of inhibitors. Corrosion damage tends to take place in low points, bends and fittings; and internal erosion damage occurs through abrasion by the product transported, typically at bends, trees, valves, etc. Erosion may cause deterioration of the inside wall and become a primary target for corrosion.

13.5.3 External Damage

- Dropped objects due to, for example, activities on or surrounding a platform,
- Abrasion between cable or chain and the pipe,
- In form of a direct hit or dragging due to anchoring; and/or
- Damage caused by construction operations, shipping operations, fishing operations.

13.5.4 Environmental Damage

- Severe storms and excessive hydrodynamic loads,
- Earthquake,
- Seabed movement and instability,
- Seabed liquefaction; and/or
- Icebergs and marine growth.

13.5.4.1 Types of Pipeline Damage

- Damage to pipe wall,
- Overstressing or fatigue damages; and/or
- Corrosion coating and weight coating damage.

13.5.5 Damage Assessment

For damaged pipes, ASME B31.4 and B31.8 may be applied to determine whether a damage assessment and repair will be necessary. If a severe damage cannot be repaired immediately, strength assessment of pipes with damages such as dents, corrosion defects and weld cracks may be performed, as defined in Chap. 15.

13.6   Pipeline Repair Methods

13.6.1 Conventional Repair Methods

Non-critical intervention work such as free-span correction, retrofitting of anode sleds and rock dumping can usually be considered as planned preventive measures. For the localised repair of nonleaking minor and intermediate pipeline damage, repair clamps may be utilised without the necessity of an emergency shutdown to the pipeline system. For major pipeline damage resulting in or likely to result in product leakage, immediate production shutdown and depressurisation is invariably required, allowing the damaged pipe section to be replaced.

13.6.2 Maintenance Repair

Non-critical repairs that in the short term will not jeopardise the safety of the pipeline and hence can form part of a planned maintenance programme. Examples are:

- Corrosion coating repair,
- Submerged weight rectification,
- Cathodic protection repair,
- Span rectification procedures, and
- Installation of an engineered backfill (rock dumping).

13.7   Extension of Use

This section pertains to obtaining and continuance of classification/certification of existing pipelines beyond the design life. The classification/certification requires special considerations with respect to the review, surveys, and strength analyses to verify the adequacy of the pipeline for its intended services. To establish if an existing pipeline is suitable for extended service, the following are to be considered:

- Review original design documentation, plans, structural modification records and survey reports,
- Survey pipeline and structures to establish condition,
- Review the results of the in-place analysis utilising results of survey, original plans, specialised geotechnical and oceanographic reports and proposed modifications which affect the dead, live, environmental and earthquake loads, if applicable, on the pipeline,
- Re-survey the pipeline utilising results from strength analysis. Make any alterations necessary for extending the service of the pipeline, and

- Review a programme of continuing surveys to assure the continued adequacy of the pipeline.

The first two items are to assess the pipeline to determine the possibility of continued use. In-place analyses may be utilised to identify the area(s) most critical for inspection at the resurvey.

Fatigue life is sensitive to the waves encountered during the past service and future prediction, and long- term environmental data is to be properly represented. Should any area be found to be deficient, then these areas require strengthening to achieve the required fatigue life. Otherwise, inspection programmes are to be developed to monitor these areas on a periodical basis.

Fatigue analysis will not be required if the following conditions are satisfied:

- The original fatigue analysis indicates that the fatigue lives of all joints are sufficient to cover the extension of use,
- The fatigue environmental data used in the original fatigue analysis remain valid or deemed to be more conservative,
- Cracks are not found during the re-survey, or damaged joints and members are being repaired, and
- Marine growth and corrosion is found to be within the allowable design limits.

Surveys, as described in Chap. 2, are to be undertaken on a periodic basis to ascertain the satisfactory condition of the pipeline. Additional surveys may be required for pipe systems having unique features.

13.7.1 Review of Design Documents

Pipeline design information is to be collected to allow an engineering assessment of a pipeline's overall structural integrity. It is essential to have the original design reports, documents and as-built plans and specifications and survey records during fabrication, installation, and past service. The Operator is to ensure that any assumptions made are reasonable and that information gathered is both accurate and representative of actual conditions at the time of the assessment. If the information cannot be provided, an assumption of lower design criteria, actual measurements or testing is to be carried out to establish a reasonable and conservative assumption.

13.7.2 Inspection

Inspection of an existing pipeline, witnessed and monitored by a Class Surveyor, is necessary to determine a base condition upon which justification of continued service can be made. Reports of previous inspection and maintenance will be reviewed, an inspection procedure developed, and a complete underwater inspection required to assure that an accurate assessment of the pipeline' condition is obtained. The corrosion protection system is to be re-evaluated to ensure that existing anodes are capable of serving the extended design life of the pipe system. If found necessary, replacement of the existing anodes or installation of additional new anodes is to be carried out. If the increase in hydrodynamic loads due to the addition of new anodes is significant, this additional load is to be account for in the strength analysis.

13.7.3 Strength Analyses

The strength analyses of an existing pipeline are to incorporate the results of the survey and any structural modifications and damages. The original fabrication materials and fit-up details are to be established such that proper material characteristics are used in the analysis and any stress concentrations are accounted for. For areas where the design is controlled by earthquake or ice conditions, the analyses for such conditions are also to be carried out. The results of the analyses are considered to be an indicator of areas needing inspection. Effects of alterations of structures or seabed to allow continued use are to be evaluated by analysis. Free spans where strength criteria are violated may be improved by seabed intervention. The results of these load reductions on the structure are to be evaluated to determine whether the repairs/ alterations are needed.

13.7.4 Implementing Repairs/Reinspection

The initial condition survey, in conjunction with structural analysis, will form the basis for determining the extent of repairs/alterations which will be necessary to class the pipeline for continued operation. A second survey may be necessary to inspect areas which the analysis results indicate as being the more highly stressed regions of the structure. Areas found overstressed are to be strengthened. Welds with low fatigue lives may be improved either by strengthening or grinding. If grinding is used, the details of the grinding are to be submitted to Class for review and approval. Intervals of future periodic surveys shall be determined based on the remaining fatigue lives of these welds.

Limit State Design Criteria

14

14.1 General

The strength requirements for pipeline design will normally be satisfied if the following limit states are fulfilled:

- Bursting,
- Local buckling and collapse,
- Fracture,
- Fatigue, and
- Ratcheting/out-of-roundness.

14.1.1 Limit States

The limit states are given in the form of maximum allowable limits such as strain, stress and bending moments and are to be checked for the following design scenarios:

- Installation,
- Empty condition,
- Water filled condition,
- Pressure test condition, and
- Operational conditions.

A. A. Olsen, *Subsea Pipeline Systems*, Synthesis Lectures on Ocean Systems Engineering, https://doi.org/10.1007/978-3-031-74790-8_14

14.1.2 Maximum Allowable Limits

The maximum allowable limits are based on equations predicting the ultimate strength to which reduction factors are applied to address uncertainties in strength estimation and consequences related to failure:

- Statistical values for design equations,
- Statistical values for material properties,
- Complexity of conditions to be modelled, and
- Location zone.

14.2   Classification of Containment, Location, Material Quality and Safety

This section recommends limit state design factors reflecting appropriate safety levels, containment hazard levels, consequences of failure related to life, environment, and business, together with uncertainties related to material resistance and loads acting on the pipe. The recommended factors will be applicable for most pipeline designs, however the guidance outlined herein is deemed inadequate, the establishment of alternative safety factors is to be verified by Class.

14.2.1 Classification of Containment

The containment being transported is to be identified and categorised according to Table 14.1.

Table 14.1 Classification of containment

Category	Description
A	Contents in the form of gases and/or liquids that at ambient temperature and atmospheric pressure are non-flammable and nontoxic Examples: water, water-based fluids, nitrogen, carbon dioxide, argon, and air
B	Contents in the form of gases and/or liquids that at ambient temperature and atmospheric pressure are flammable, toxic and/or could lead to environmental pollution if released Examples: oil, petroleum products, toxic liquids, hydrogen, natural gas, ethane, ethylene, ammonia, chlorine, liquefied petroleum gas (such as propane and butane), natural gas liquids, etc.

14.2.2 Classification of Location

The pipeline system is to be classified into location classes, as defined in Table 14.2.

The material quality used for line pipes designed based on the format given in this chapter is expected to meet the requirements of chemical composition and mechanical properties as defined by ISO 3183 1–3. The material quality is to be classified as A, B or C, as given by Table 14.3.

The tensile and Charpy V-notch impact properties are to be in accordance with relevant specifications, such as API SPEC 5L or ISO 3183 1–3. Tensile properties of the line pipe are to be tested in both transverse and longitudinal directions, while Charpy V-notch samples are to be tested only in the transverse direction. Materials are to exhibit fracture toughness that is satisfactory for the intended application, as supported by previous satisfactory service experience or appropriate toughness tests. Where the presence of ice is judged as a significant environmental factor, material selection may require special consideration.

The following requirements with respect to the definition of Specified Minimum Yield Strength and Specified Minimum Tensile Strength in the circumferential direction are to be fulfilled:

$$SMYS < (mean - 2 * Standard\ Deviations)\ of\ yield\ stress$$

$$SMTS < (mean - 3 * Standard\ Deviations)\ of\ tensile\ stress$$

The material yield stress and tensile stress are to be obtained through testing at relevant temperatures and test results submitted to Class. Material anisotropy is defined by the

Table 14.2 Classification of location

Zone	Description
1	Areas where infrequent human activity is anticipated (to be documented)
2	Areas adjacent to manned platforms and areas where frequent human activity is anticipated

Table 14.3 Classification of material quality

Material quality	Description
Class A	A basic quality level corresponding to that specified in the main part of API SPEC 5L or ISO 3183-1
Class B	For transmission pipelines, overall enhanced requirements (e.g., concerning toughness and non-destructive testing) are addressed, as specified by ISO 3183-2
Class C	For particularly demanding applications where very stringent requirements (e.g., concerning sour service, fracture arrest properties, plastic deformation) on quality and testing are imposed, ISO 3183-3

ratio between material properties in the transverse direction and those in the longitudinal direction, and is, where appropriate, to be accounted for in the strength calculations. For pipelines or sections of these to be operated at temperatures above 50 °C (120 °F), appropriate material resistance de-rating factors are to be established and applied to the specified minimum yield and tensile strength. For line pipes manufactured by the UO or UOE method, the influence of the manufacturing process on the yield stress is to be accounted for in the design. If the material's specified minimum yield stress is less than 70 ksi, the values given in Chap. 4 may be used, otherwise, the influence of the manufacturing process is to be based on testing.

14.2.3 Classification of Safety

The definition of safety classes used in this chapter are in accordance with Table 14.4.

For normal use, the safety classes in Tables 14.5 and 14.6 apply. Other safety classification may be justified based on the design target reliability level of the pipeline but are in such cases to be submitted for approval by Class.

14.3   Hoop Stress Criteria

The hoop stress is not to exceed the following:

Table 14.4 Definition of safety classes

Safety class	Description
Low	Failure implies no risk to human safety and only limited environmental damage and economic losses
Medium	Failure implies negligible risk to human safety and only minor damage to the environment but may imply certain economic losses
High	Failure implies risk to the total safety of the system so as to human safety and environmental pollution. High economic losses may apply

Table 14.5 Classification of safety classes pipelines

Load condition	Content category A		Content category B	
	Location zone 1	Location zone 2	Location zone 1	Location zone 2
Temporary	Low	Low	Low	Low
Operational	Low	Normal	Normal	High
Abnormal	Low	Normal	Low	Normal

Table 14.6 Classification of safety classes for bundles

Pipes	Launch and installation	Operation
Flowlines	Low	Normal
Heat-up pipes	Low	Low
Carrier pipe and sleeve	Normal	Low

Table 14.7 Usage factors for hoop stress criteria

Material quality	Usage factor	Safety class		
		Low	Normal	High
Class B, C	η_s	0.85	0.80	0.70
Class A	η_s	0.83	0.77	0.67

$$(p_i - p_e)\frac{D - t}{2 \cdot t} \leq \eta_s \cdot \min[1.00 \cdot SMYS, \ 0.87 \cdot SMTS]$$

where:

p_i	internal pressure
p_e	external pressure
D	nominal outside steel diameter of pipe
t	minimum
$SMYS$	Specified Minimum Yield Strength at design temperature
$SMTS$	Specified Minimum Tensile Strength at design temperature

Derating of material resistance is, where applicable, to be accounted for in the definition of Specified Minimum Yield Strength and Specified Minimum Tensile Strength at elevated design temperatures. For thick-walled pipes with a $D/t < 20$, the above hoop stress criteria may be adjusted based on, e.g., BS 8010-3. The usage factors for hoop stress criteria are given in Table 14.7.

14.3.1 System Pressure Test

The pipeline system is to be subjected to a system pressure test after installation in accordance with Chap. 12.

14.4   Limit State for Local Buckling

Local buckling may occur due to excessive bending combined pressure and longitudinal force. The failure mode will be a combination of local yielding and flattening/local buckling and mainly depends on the diameter to wall thickness ratio, load condition and local imperfections in material and geometry. Initial out-of-roundness is the only local imperfection accounted for in the equations presented in this chapter. The formulas in this Section are applicable to diameter to thickness ratios between 10 and 60. For pipes of larger D/t ratio, the use of the moment criterion is subject to Class approval. The criteria will be applicable to steel pipes, but may be applied to titanium pipes, provided that an equivalent $SMYS$ is defined as the minimum of $SMYS$ and $SMTS/1.3$.

14.4.1 Maximum Allowable Moment

The maximum allowable bending moment, M_{All}, ensuring structural strength against local buckling may be found by:

$$M_{All} = \frac{\eta_{RM}}{\gamma_c} M_\ell \cdot \sqrt{1 - \left(1 - \alpha^2\right) \cdot \left(\frac{p}{\eta_{RP}p_\ell}\right)^2}$$

$$\cdot \cos\left[\frac{\pi}{2} \cdot \frac{\frac{\gamma_c F}{\eta_{RF} F_\ell} - \alpha \cdot \frac{p}{\mu_{RP}p_\ell}}{\sqrt{1 - \left(1 - \alpha^2\right) \cdot \left(\frac{p}{\eta_{RP}p_\ell}\right)^2}}\right]$$

where:

p pressure acting on the pipe $(p_i - p_e)$
F true longitudinal force acting on the pipe
γ_c condition load factor
η_R strength usage factors

The moment M_ℓ which is the moment capacity in pure bending, may be calculated as:

$$M_\ell = \left(1.05 - 0.0015 \cdot \frac{D}{t}\right) \cdot SMYS \cdot D^2 \cdot t$$

where:

$SMYS$ Specified Minimum Yield Strength in longitudinal direction at design temperature
D average diameter

t wall thickness

The longitudinal force, F_ℓ, may be estimated as:

$$F_\ell = 0.5(SMYS + SMTS)A$$

where:

A cross sectional area, which may be calculated as $\pi \times D \times t$
$SMYS$ Specified Minimum Yield Strength in longitudinal direction at design tempera-
         ture
$SMTS$ Specified Minimum Tensile Strength in longitudinal direction at design temper-
         ature

The pressure, p_ℓ, is for external overpressure conditions equal to the pipe collapse pressure and may be calculated based on:

$$p_\ell^3 - p_{el} \cdot p_\ell^2 - \left(p_p^2 + p_{el} \cdot p_p \cdot f_0 \cdot \frac{D}{t} \right) \cdot p_\ell + p_{el} \cdot p_p^2 = 0$$

where:

p_{el} $\frac{2 \cdot E}{(1-v^2)} \cdot \left(\frac{t}{D} \right)^3$
p_p $k_{fab} \cdot SMYS \cdot \frac{2 \cdot t}{D}$
D average diameter
t wall thickness
f_0 initial out-of-roundness $(D_{max}-D_{min})/D$, not to be taken less than 0.5%

Note: out-of-roundness caused during the construction phase is to be included but not flattening due to external water pressure or bending in as-laid position. Increased out-of-roundness due to installation and cyclic operating loads may aggravate local buckling and is to be considered. It is recommended that out-of-roundness due to through-life cyclic loads be simulated, if applicable.

$SMYS$ Specified Minimum Yield Strength in hoop direction at design temperature
E Young's Modulus at design temperature
v Poisson's ratio
k_{fab} material resistance de-rating factor due to fabrication

For internal overpressure conditions, the pressure, p_ℓ, is equal to the burst pressure, which may be found as:

$$p_\ell = 0.5 \cdot (SMTS + SMYS) \cdot \frac{2 \cdot t}{D - t}$$

where:

SMYS   Specified Minimum Yield Strength in hoop direction at design temperature
SMTS   Specified Minimum Tensile Strength in hoop direction at design temperature
D   average diameter
t   wall thickness

The strength anisotropy factor, α, may be calculated as:

$$\alpha = \frac{\pi \cdot D^2}{4} \cdot \left| \frac{p_c}{F_\ell} \right| \quad \text{for external overpressure}$$

$$\alpha = \frac{\pi \cdot D^2}{4} \cdot \left| \frac{p_b}{F_\ell} \right| \quad \text{for internal overpressure}$$

14.5   Effects of Manufacturing Process

The material strength in the hoop direction will be influenced by the manufacturing process, and if no test data are available for the hoop strength, the following reduction factor, k_{fab}, is to be used.

$$k_{fab} \begin{cases} 1.00 \text{ Seamless and annealed pipes} \\ 0.93 \text{ Welded pipes not expanded, e.g., UO pipes} \\ 0.85 \text{ Welded and expanded pipes, e.g., UOE pipes} \end{cases}$$

14.6   Usage Factors

Usage factors, η_R, are listed in Table 14.8.

Table 14.8 Usage factors

Usage Factors	η_{RP}	η_{RF}	η_{RM}
Low	0.95	0.90	0.80
Normal	0.93	0.85	0.73
High	0.90	0.80	0.65

Load condition	Condition load factor γ_C
Uneven seabed	1.06
Continuously stiff supported	0.85
System pressure test	0.94
Otherwise	1.0

Table 14.9 Condition load factors for limit state design

14.7   Condition Load Factors

To account for uncertainties related to the modelling of certain load conditions, condition load factors are introduced in accordance with Table 14.9.

14.8   Limit State for Fracture of Girth Weld Crack-Like Defects

Fracture in welds due to tensile strain is normally evaluated in accordance with a recognised assessment method based on the failure assessment diagram, which combines the two potential failure modes, brittle fracture, and plastic collapse. This section may be applied to define acceptance criteria for inspection of girth welds.

14.8.1   Possible Cracks in Girth Weld

Various types of imperfections are known to occur in girth welds. The most damaging types are cracks, inadequate penetration of the root bead and lack of fusion. The imperfections are particularly damaging if they occur in a weld that significantly under-matches the yield strength of the base material. When evaluating the risk of fracture failure for girth welds, assumptions are to be made regarding types, dimensions, and locations of weld defects. The most frequently seen type of planar/crack-like defect in one-sided Shielded Metal Arc Welding is lack-of-fusion defects. Such defects can be located near the surface, at the root of the weld toe or be surface-breaking and may have gone undetected when following non-destructive testing procedures according to API STD 1104.

Maximum weld flaws are to be used as the basic input for the assessment. The flaw may be assumed as maximum allowable defect due to lack of fusion between passes. Surface flaw is chosen as the worst-case scenario from acceptable flaws specified in the welding procedure specifications. The defect sizes to be used in the fracture assessment are to be based on the welding methods used and the accuracy of the non-destructive testing during construction. If no detailed information is available, the defects and material may be taken as below:

Type	Surface flaw due to lack of fusion
Depth (a)	Minimum of 3 mm (0.118 inch) and nominal wall-thickness divided by number of welding passes
Length (2c)	2c < min [2t, 2 50 mm (1.97 inch)]
CTOD	0.075 mm (0.003 inch) or as given by welding specifications
Material	As for parent material
D	average diameter
t	wall thickness

The fracture failure of welds is highly dependent on the weld matching and on the ratio of yield to tensile strength. An adequate strain concentration factor is to be established, accounting for the stiffness of coatings and buckle arrestors.

14.8.2 Fracture of Cracked Girth Welds

Defects in girth welds can be assessed on one of three levels, depending upon the quality of the affected welds, the availability of relevant material data and difficulties related to repairs. Level 1 is mandatory, where acceptance levels are graded according to criticality of application. Level 2 is a conservative initial assessment. Level 3 should be applied to details that fail in Level 2.

14.8.2.1 Level 1 Assessment—Workmanship Standards

Pipeline welding codes establish minimum weld quality standards based on inspection of welder's workmanship, and the flaw acceptance criteria are evolved through industry experience. Hence, most workmanship standards are similar, though not identical, in terms of allowable imperfection types and sizes. The workmanship standards may be based on API STD 1104 or standards such as ASME Boiler and Pressure Vessel Code, CSA Z662 and BS 4515.

14.8.2.2 Level 2 Assessment—Alternative Acceptance Standards

Alternative acceptance standards have been developed to facilitate acceptance of flaws that do not meet workmanship standards. Incentives for alternative standards are usually economic, arising due to the inaccessibility or quantity of welds that would otherwise be repaired. Alternative standards recognise that the true severity of a flaw is dependent on material toughness and applied stress levels and can only be determined using fracture mechanics principles.

CTOD is established from destructive tests performed on weldments. If the pipeline is yet to be constructed, CTOD tests can be performed as part of the weld procedure qualification. If the pipeline is already in service and CTOD data are not available, the welding procedures, consumable and base materials used in construction may be used to duplicate welds for the purpose of conducting CTOD tests. If any of these elements is no longer available, it will be necessary to obtain a representative weld for testing. Alternatively, a lower-bound CTOD of 0.075 mm (0.003 inch) may be assumed in lieu of tests. Alternative criteria are given in codes and standards, such as in Appendix to API STD 1104, Appendix K to CSA Z662, BSI 7910 and the EPRG Guidelines on assessment of defects in transmission pipeline girth welds.

14.8.2.3   Level 3 Assessment—Detailed Analysis

A more detailed analysis is to use FADs and tearing stability analysis from BS 7910, API RP 579 or equivalent. BS 7910 Level 2 provides three possible FADs:

(1) Generalised curve for low work hardening materials,
(2) Generalised curve for low and high work hardening materials, and
(3) Material specific curve.

Level 3 in BS 7910 covers tearing instability analysis.

In the assessment of the fracture failure capacity of the weld due to longitudinal strain of the pipe, surface breaking flaws may be idealised as semi-elliptical surface weld defects of depth, a, and total length, 2c. a and c are to be based on non-destructive test measurements with a minimum of not less than the minimum detectable crack size by the applied non-destructive test method. Parametric solutions for K are available in codes such as API RP 579 and BS 7910.

The critical stress levels with respect to failure (e.g., fracture, plastic collapse of remaining ligament) can be obtained from Failure Assessment Diagram analysis, for the different pipelines as a function of the degree of corrosion wall-thickness reduction. The corresponding critical strain level is estimated using the Ramberg–Osgood curve for the stress–strain relationship.

14.8.3   Fatigue Crack Propagation

The fatigue crack propagation rate may be obtained following the procedure described in Chapter 14 or refer to basic fracture mechanics methodology as per BS 7910, API RP 579 or equivalent. Use actual da/dN versus ΔK and ΔK data wherever possible. Fit Paris' equation over the entire range of data or fit alternative crack growth law (e.g., Forman equation, see API RP 579 for others). Alternatively, fit Paris' equation in piecewise linear manner. The latter may be the most effective for crack growth in seawater with/without cathodic protection. In the absence of specific da/dN data, use upper bound relationships in BS 7910, API RP 579 or equivalent. Fatigue life versus crack depth may be predicted by integrating $\Delta a/\Delta N$ versus ΔK relationship using weight average ΔK or cycle-by-cycle approach if interaction effects are negligible.

14.9 Limit State for Fatigue

Unsupported pipeline spans, welds, J-lay collars, and buckle arrestors are to be assessed for fatigue. Potential cyclic loading that can cause fatigue damage includes vortex-induced vibrations, wave induced hydrodynamic loads, floating installation movements and cyclic pressure and thermal expansion loads. The fatigue life is defined as the time it takes to develop a through-wall-thickness crack. Fatigue analysis and design can be conducted using:

(1) S–N approach for high cycle fatigue (and low cycle fatigue of girth welds may be checked based on $\Delta\varepsilon - N$ curves),
(2) Fracture mechanics approach, and/or
(3) Hybrid approach (combination of S–N and fracture mechanics approaches).

It is recommended that the S–N approach be prioritised for design purposes. The fracture mechanics approach is recommended for assessments or reassessments and establishment of inspection criteria. It should be noted that the S–N approach and fracture mechanics approach should produce very similar results. Different acceptance criteria could be applied for the different approaches. In general, design life criteria should be adopted when using the S–N approach, while through-thickness crack criteria are to be applied for the fracture mechanic approach. To achieve a consistent safety level, it might be necessary to calibrate the initial crack size in the fracture mechanics approach against the SN approach.

14.9.1 Fatigue Assessment Based on S–N Curves

For assessment of high cycle fatigue, fatigue strength is to be calculated based on laboratory tests (S–N curves) or fracture mechanics. In the limit state-based fatigue analysis, appropriate partial safety factors are to be defined and applied to loads and material strength prior to the estimation of the accumulated fatigue damage. Typical steps required for fatigue analysis using the S–N approach are outlined below.

(1) Estimate long-term stress range distribution,
(2) Select appropriate S–N curve,
(3) Determine stress concentration factor, and
(4) Estimate accumulated fatigue damage using Palmgren–Miner's rule.

The S–N curves to be used for fatigue life calculation may be defined by the following formula:

$$\log N - \log a - m \cdot \log \Delta\sigma$$

where:

N allowable stress cycle numbers

a, m parameters defining the curves, which are dependent on the material and structural detail

$\Delta\sigma$ stress range, including the effect of stress concentration

Multiple linear S–N curves in logarithmic scale may be applicable. Each linear curve may then be expressed as the above equation. The fatigue damage may be based on the accumulation law by Palmgren–Miner:

$$D_{fat} = \sum_{i=1}^{M_C} \frac{n_i}{N_i} \leq \eta$$

where:

D_{fat} accumulated fatigue damage

η allowable damage ratio

N_i number of cycles to failure at the ith stress range defined by the S–N curve

n_i number of stress cycles with stress range in block i

14.9.2 Fatigue Assessment Based on $\Delta\varepsilon$ – N Curves

The number of strain cycles to failure may be assessed according to the American Welding Society (AWS) Standards $\Delta\varepsilon$ – N curves, where N is a function of the range of cyclic bending strains $\Delta\varepsilon$. The strain range $\Delta\varepsilon$ is the total amplitude of strain variation, i.e., the maximum less the minimum strains occurring in the pipe body near the weld during steady cyclic bending loads.

14.9.3 Fatigue Assessment Based on Fracture Mechanics

In the fracture mechanics approach, the crack growth is calculated using Paris' equation and the final fracture found in accordance with recognised failure assessment diagrams. It may be applied to develop cracked S–N curves for pipes containing initial defects. If a crack growth analysis is performed by the fracture mechanics method, the design criterion for fatigue life is to be at least 10 times the service life for all components. The initial flaw size is to be the maximum acceptable flaw specified for the non-destructive testing during manufacture of the component in question. Typical steps required for fatigue analysis using the fracture mechanics approach are outlined below:

(1) Estimate long-term stress range distribution,
(2) Determine stress concentration factor,
(3) Select appropriate material parameter to be used in Paris' equation,
(4) Determine initial crack size and crack initiation time,
(5) Determine the critical crack size,
(6) Determine geometry function for stress intensity factor, and
(7) Integrate Paris' equation to estimate fatigue life.

14.10 Limit State for Ratcheting/Out-of Roundness

The pipeline out-of-roundness is related to the maximum and minimum pipe diameters (D_{max} and D_{min}) measured from different positions around the sectional circumference according to:

$$f_0 = \frac{D_{max} - D_{min}}{D}$$

Out-of-roundness introduced during manufacturing, storage and transportation is generally not to exceed 0.75%. For design, the initial out-of-roundness is not to be assumed less than 0.50%. The out-of-roundness of the pipe may increase where the pipe is subject to reverse bending and the effect of this on subsequent straining is to be considered. For a typical pipeline, the following scenarios will influence the out-of-roundness:

(1) Reverse inelastic bending during installation, and
(2) Cyclic bending due to shutdowns in operation if global buckling is allowed to relieve temperature- and pressure-induced compressive forces.

Critical point loads may arise at free-span shoulders, artificial supports, and support settlement. Out-of-roundness accumulative through the life cycle is, if applicable, to be found from ratcheting analysis. Out-of-roundness is not to exceed 2% unless:

(1) Effect of out-of-roundness on moment capacity and strain criteria is included, and
(2) Requirements for pigging and other pipe run tools are met.

Ratcheting is described in general terms as signifying incremental plastic deformation under cyclic loads in pipelines subject to high pressure and high temperatures. The effect of ratcheting on out-of-roundness and local buckling is to be considered. A simplified code check of ratcheting is that the equivalent plastic strain is not to exceed 0.1%, as defined based on elastic-perfectly-plastic material and assuming that the reference state for zero strain is the as-built state after mill pressure testing. The finite element method may be applied to quantify the amount of deformation induced by ratcheting during the life cycle of a pipeline.

14.11   Finite Element Analysis of Local Strength

If adequate documentation is presented, alternative methods for estimating the strength capacity of pipes subjected to combined loads may be accepted for installation and seabed intervention design. An important tool is the finite element method, which allows the designer to model the geometry, material properties and imperfections such as out-of-roundness, field joints, attachments and corrosion defects and thereafter estimate the maximum strength for a given load scenario. Another important issue is the effect of cyclic loads. Cyclic loads may aggravate/increase imperfections and fatigue damage, which may end up reducing the design life.

14.11.1   Modelling of Geometry and Boundary Conditions

When applicable, the size of a finite element model may be reduced by introducing symmetry boundary conditions. For this approach to be valid, material, geometry and loads are to be symmetrically distributed around the same symmetry line. This approach may reduce finite element models to one half or less of the pipe section. It is important that the modelled pipe section includes all features and attachments relevant for the stress distribution in the pipe and that the model is sufficiently long to catch relevant failure modes.

Imperfections such as out-of-roundness, weld defects/misalignment and corrosion defects may reduce the strength of the pipe considerably, and the largest realistic imperfections are to be included in the model. Imperfections that might be insignificant for some load conditions will be catastrophic for others. An example is out-of-roundness which has negligible influence on the burst strength while it might reduce the collapse strength due to considerable external overpressure. If different imperfections are combined, their mutual orientation might influence the pipe strength. An example is combined out-of-roundness and corrosion defects. Here, the worst orientation of the out-of-roundness with respect to pipe strength might change with the increasing size of a corrosion defect.

14.11.2 Mesh Density

The mesh density generally depends on the selected element definition and detail level of the analysis. Selection of element definition and mesh density/distribution is generally to be based on engineering experience, but it is recommended that sensitivity studies be performed to demonstrate the adequacy of the chosen mesh. This exercise will normally be a part of the verification procedure for the model.

14.11.3 Material

The material input to Finite Element Software is to be based on the software used and the problem to be solved. For pure elastic analysis, it will normally be enough to describe the material characteristic in the form of the modulus of elasticity and Poisson's ratio, while for analysis, including material nonlinearity, it will be necessary to include a description of the material's plastic behaviour. Here, it is important to notice that finite element programmes might use either a true stress–strain or engineering stress–strain relationship. It is important to consider if residual stresses in the material are to be included in the model. For pipelines, residual stresses in both longitudinal and hoop directions might be introduced through the forming process or by seam welding. Effects of these residual stresses are, in general, considered negligible for pipeline analysis, but practice has demonstrated that this is not always the case. When cyclic inelastic loading is to be modelled, a nonlinear combined isotropic and kinematic material model may be applied. This material model will also be applicable for low cycle fatigue studies.

14.11.4 Load and Load Sequence

The sequence of applied loads may influence the result, and if the sequence is not known, several tests are to be performed to make sure that the results represent the worst case. As an example, a pipe subjected first to axial tension and then pressure might burst at a higher pressure than for the pure pressure condition, and vice versa.

14.11.5 Validation

Finite element models and other analysis models, in general, are to be validated against appropriate mechanical tests and the validation approved by Class before they are used in design.

Assessment of Corrosion, Dents and Crack-Like Defects

15

15.1 General

This chapter defines strength criteria that may be used for corrosion allowance design and assessment of corroded and dented pipes and girth weld defects.

15.2 Scope of the Assessment

The scope of the assessment includes:

(1) Proper characterisation of defects by thickness profile measurements,
(2) An initial screening phase to decide whether detailed analysis is required or whether fitness for service is to be considered,
(3) Detailed assessment phase:
   - Check burst limit state (allowable versus maximum internal service pressure),
   - Check collapse limit state (allowable versus maximum external service pressure, bending moment and axial load),
   - Check adequacy of residual corrosion allowance for remaining service life,
   - Other checks, e.g., residual fatigue life, particularly if cracks are detected within defects, and
(4) Updated inspection and maintenance programme.

© The Author(s), under exclusive license to Springer Nature Switzerland AG 2025
A. A. Olsen, *Subsea Pipeline Systems*, Synthesis Lectures on Ocean Systems
Engineering, https://doi.org/10.1007/978-3-031-74790-8_15

15.3   Corrosion Defect Inspection

The pipe is to be inspected, if applicable, for defects due to corrosion, and consideration is to be given to the uncertainties in corrosion rate and measurement accuracy. These uncertainties may be modelled by a probabilistic method. Initially, inspection intervals may be set as three (3) years for oil and water lines and six (6) years for dry gas lines. However, the optimum inspection intervals may be selected using reliability-based inspection planning techniques. The reliability-based inspection planning procedure is to estimate the corrosion rate in the line and identify all the defects that would fail the strength requirements before the next inspection.

15.4   Corrosion Defect Measurements

The actual size and shape of the corrosion defect is to be defined by an adequate number of measured thickness profiles. These measurements are to be performed in accordance with Sect. 5.3 of API RP 579 or equivalent. The assessment of a single isolated defect is to be based on a critical profile defined by the largest measured characteristic dimensions of the defect (e.g., depth, width, length) and properly calibrated safety/uncertainty factors to account for uncertainties in the assessment and thickness measurements. Refer to API RP 579 Sect. 4.3 for guidance on extracting the critical profile from measured thickness profiles. If several corrosion defects inside a relatively small area have been detected during an inspection, appropriate determination of defect interaction is to be conducted, accounting for the following factors:

- Angular position of each defect around circumference of the pipe,
- Axial spacing between adjacent defects,
- Internal or external defects,
- Length of individual defects,
- Depth of defects, and
- Width of defects.

A distance equivalent to the nominal pipe wall thickness may be used as a simple criterion of separation for colonies of longitudinally oriented pits separated by a longitudinal distance or parallel longitudinal pits separated by a circumferential distance. For longitudinal grooves inclined to pipe axis:

- If the distance x, between two longitudinal grooves of length L_1 and L_2, is greater than either of L_1 or L_2, then the length of corrosion defect L is L_1 or L_2, whichever is greater. It can be assumed that there is no interaction between the two defects; or

- If the distance x, between two longitudinal grooves of length L_1 and L_2, is less either of L_1 and L_2, it is to be assumed that the two defects are fully interacted and the length of the corrosion defect L is to be taken as $L = L_1 + L_2 + x$.

15.5 Corrosion Defect Growth

The corrosion defect depth, d, after the time of operation, T, may be estimated using an average corrosion rate V_{CR}:

$$d = d_0 + V_{CR} \cdot T$$

where d_0 is defect depth at the present time. The defect length may be assumed to grow in proportion with the depth, hence:

$$L = L_0 \left(1 + \frac{V_{CR} \cdot T}{d_0}\right)$$

where L and L_0 are defect lengths at the present time and the time T later. The corrosion rate is to be based on relevant service data or laboratory tests.

15.6 CO₂ Corrosion Rate Estimate

CO_2 corrosion rates in pipelines made of carbon steel may be evaluated using industry-accepted equations that preferably combine contributions from flow independent kinetics of the corrosion reaction at the metal surface, with the contribution from flow dependent mass transfer of dissolved CO_2. The corrosion rate V_{CR}, in mm/year, can be predicted by:

$$V_{CR} = \frac{1}{\frac{1}{V_r} + \frac{1}{V_m}}$$

where:
$V_r =$ flow independent contribution, denoted the reaction rate.

The reaction rate V_r can be approximated by:

$$log(V_r) = 4.93 - \frac{1119}{T_{mp} + 273} + 0.58 \cdot log(pCO_2)$$

where:
$T_{mp} =$ temperature, in °C.
$pCO_2 =$ partial pressure of CO_2, in bar.
$nCO_2 \cdot p_{opr} = CO_2$ operating pressure in bar.

nCO_2 = fraction of CO_2 in the gas phase.
p_{opr} = operating pressure, in bar.

The mass transfer rate v_m is approximated by:

$$v_m = 2.45 \frac{U^{0.8}}{d^{0.2}} \cdot pCO_2$$

where:
U = liquid flow velocity, in m/s.
d = inner diameter in metres.

15.7 Maximum Allowable Operating Pressure for Corroded Pipes

All defects deeper than 0.8 times the wall thickness must be repaired. For less severe defect depths, ASME B31G or the following criteria may be applied to estimate the maximum allowable operating pressure for pipelines with a single corrosion defect (Table 15.1):

$$MAOP \leq \eta \cdot p_{burst} = \eta \cdot 0.5 (SMYS + SMTS) \left(\frac{2t}{D} \right) \frac{1 - \left(\frac{d}{t} \right)}{1 - \frac{d}{t\sqrt{1 + 0.8 \left(L/\sqrt{Dt} \right)^2}}}$$

where:
$MAOP$ = maximum allowable operating pressure.
p_{burst} = internal overpressure at burst.
    D = average diameter.
t = wall thickness measurement.
d = depth of corrosion defect, not to exceed $0.8 \times t$
L = length of corrosion defect.

Table 15.1 Usage factors for hoop stress criteria

Usage factor	Safety class		
	Low	Normal	High
η	0.76	0.72	0.63

Note
[1] Safety classes are in accordance with Chap. 14
[2] Load factors and condition load factors are in accordance with Chap. 14

$SMYS$ = Specified Minimum Yield Strength in hoop direction.
$SMTS$ = Specified Minimum Tensile Strength in hoop direction.
η = usage factor which is to be taken in accordance with.

15.8   Maximum Allowable Moment

Local buckling may occur due to excessive combined pressure, longitudinal force and bending. The failure mode will be a combination of local yielding and flattening/local buckling. The failure mode mainly depends on the diameter to wall thickness ratio (D/t), load condition and local imperfections in material and geometry. This section gives the maximum allowable bending moment for pipes with a local corrosion defect. The formulas in this chapter are applicable to diameter to thickness ratio in between 10 and 60. For pipes of larger D/t ratio, the use of the moment criterion is subjected to Class approval. The maximum allowable bending moment for local buckling of corroded pipes, valid for both internal- and external- overpressure, can be expressed as given below. The bending strength is to be found by both calculations 1 and 2, after which the lowest value is to be used as the maximum allowable design moment. The maximum allowable bending moment for local buckling can be found by:

$$M_{all} = \frac{\eta_{RM}}{\gamma C} M_\ell \left[\delta_1 \sin(\varphi) \sqrt{1 - (-\alpha^2)\left(\frac{p}{\eta_{RP}P_\ell}\right)^2} + 0.5(1 - k_2)\sin(\beta)\alpha \frac{p}{\eta_{RP}|P_\ell|} \right.$$

$$\left. + \delta_2 \sqrt{1 - (1 - \alpha^2)\left(\frac{p}{\eta_{RP}P_\ell}\right)^2} \right]$$

where the angle to the plastic neutral axis is given by:

$$\varphi\frac{\pi + \delta_3(1 - k_1)\beta}{2\delta_4} + \frac{\pi + = -(1 - k_1)\beta}{2\delta_5}\Delta, \quad \Delta = \frac{\left(\frac{\gamma C^F}{\eta_{RP}|F_\ell|}\right) - \alpha\left(\frac{p}{\eta_{RP}|P_\ell|}\right)}{\sqrt{1 - (1 - \alpha^2)\left(\frac{p}{\eta_{RP}P_\ell}\right)^2}}$$

For grooves on the inner side of the pipe wall:

$$k_1 = \left(1 - \frac{d}{t}\right)\left(1 + \frac{d}{D}\right), k_2\left(1 - \frac{d}{t}\right)\left(1 + \frac{d}{D}\right)^2$$

For grooves on the outer side of the pipe wall:

$$k_1 = \left(1 - \frac{d}{t}\right)\left(1 - \frac{d}{D}\right), k_2\left(1 - \frac{d}{t}\right)\left(1 - \frac{d}{D}\right)^2$$

The strength anisotropy factor is given by:

$$\alpha = \frac{\pi \cdot D^2}{4} \cdot \left| \frac{P_\ell}{F_\ell} \right|$$

For extreme pressure and/or axial force conditions, it is advised that this factor be verified by finite element analysis.

Calculation 1

If $\beta \geq \frac{\pi}{1+k_1} \frac{(1-\Delta)}{(1+\Delta)}$

then $\delta_1 = 1$, $\delta_2 = 1$, $\delta_3 = 1$, $\delta_4 = 1$, $\delta_5 = 1$,

else $\delta_1 = k_1$, $\delta_2 = k_1$, $\delta_3 = k_1$, $\delta_4 = k_1$, $\delta_5 = k_1$

Calculation 2

If $\beta \geq \frac{\pi}{1+k_1} \frac{(1+\Delta)}{(1-\Delta)}$

then $\delta_1 = -1$, $\delta_2 = 1$, $\delta_3 = 1$, $\delta_4 = 1$, $\delta_5 = 1$,

else $\delta_1 = -k_1$, $\delta_2 = -1$, $\delta_3 = -1$, $\delta_4 = k_1$, $\delta_5 = k_1$

where:
M_{all} = allowable bending moment
p = pressure acting on the pipe
F = longitudinal force acting on the pipe
D = average diameter
t = wall thickness
d = defect depth
φ = angle from bending plane to plastic neutral axis
k_i = constant
α = strength anisotropy factor
β = half the defect width
δ_i = constants
γ_C = condition load factor
η_R = strength usage factor, in accordance with this chapter.

The moment M_ℓ, which is the moment capacity in pure bending, may be calculated as:

$$M_\ell = \left(1.05 - 0.0015 \cdot \frac{D}{t} \right) \cdot SMYS \cdot D^2 \cdot t$$

where:
D = average diameter

$t =$ wall thickness

$SMYS =$ Specified Minimum Yield Strength in longitudinal direction

The longitudinal force F_ℓ may be estimated as:

$$F_\ell = 0.5(SMYS + SMTS)[\pi - (1 - k_1)\beta]Dt$$

where:

$D =$ average diameter

$t =$ wall thickness

$\beta =$ half of the corrosion defect angle

$SMYS =$ Specified Minimum Yield Strength in longitudinal direction

$SMTS =$ Specified Minimum Tensile Strength in longitudinal direction

$k_1 =$ constant given in this chapter.

The pressure $P_{e\ell}$ is for external overpressure conditions equal to the pipe collapse pressure and may be calculated based on:

$$p_\ell^3 - p_{e\ell}p_\ell^2 - \left[p_p^2 + p_{e\ell}p_pf_0\left(\frac{D}{t_{cor}}\right)\right]p_\ell + p_{e\ell}p_p^2 = 0$$

where:

$p_{e\ell} = \frac{2E}{(1-v^2)}\left(\frac{t_{cor}}{D}\right)^3$

$p_p = \eta_{fab}SMYS\frac{2t_{cor}}{D}$

$t_{cor} =$ minimum measured wall thickness, including proper reduction factor accounting for inaccuracy in measurement method

$f_0 =$ initial out-of-roundness, $(D_{max} - D_{min})/D$

Note: Out-of-roundness caused during the construction phase is to be included, but not flattening due to external water pressure or bending in as-laid position. Increased out-of-roundness due to installation and cyclic operating loads may aggravate local buckling and are to be considered. Here, it is recommended that out-of-roundness due to through life loads be simulated using finite element analysis.

$SMYS=$Specified Minimum Yield Strength in hoop direction

$E =$ Young's modulus

$v =$ Poisson's ratio

$k_{fab} =$ fabrication de-rating factor given in Chap. 7.

For internal overpressure conditions, the pressure p_ℓ is equal to the burst pressure, which may be found as:

$$p_\ell = 0.5(SMYS + SMTS)\left(\frac{2t}{D}\right)\frac{1-\frac{d}{t}}{1 - \frac{d}{t\sqrt{1+0.8\left(L/\sqrt{Dt}\right)^2}}}$$

where:
D = average diameter
t = wall thickness measurement
d = depth of corrosion defect, not to exceed $0.8 \times t$
L = measured length of corrosion depth
$SMYS$ = Specified Minimum Yield Strength in hoop direction
$SMTS$ = Specified Minimum Tensile Strength in hoop direction

Load factors and condition load factors are to be taken in accordance with Chap. 14, and resistance factors η_R are listed in Chap. 14, Table 1.

15.9 Maximum Allowable Operating Pressure for Dented Pipes

The determination of maximum allowable operating pressure for a dented pipe with cracks may be calculated as:

$$P = 2 \cdot \sigma \cdot \frac{t}{D} \cdot \eta$$

where the usage factor η may be taken as 0.72, and the critical stress at failure σ is given by:

$$\sigma = \frac{2 \cdot \sigma_p}{\pi} \cdot \cos^{-1}\left[\exp\left(-\frac{\pi \cdot k_{mat}^2}{\gamma^2 \cdot 8 \cdot \alpha \cdot \sigma_p^2}\right)\right]$$

The plastic failure stress σ_p of may be taken as:

$$\sigma_p = \sigma_f \cdot \frac{t - \alpha}{t - \dfrac{\alpha}{\sqrt{1 + 0.8\left(L/\sqrt{Dt}\right)^2}}}$$

where:
D = average pipe diameter
t = pipe wall thickness
L = length of dent
α = maximum depth of pipe wall thickness defect
σ_f = flow stress, here defined as the mean value of Specified Minimum Yield and Tensile Strength
    Pipe toughness is measured in terms of the Charpy energy C_v. This measure is a qualitative measure for pipe toughness and has no theoretical relation with the fracture toughness parameter, K_{mat}. Therefore, it is necessary to use an empirical relationship between K_{mat} and C_v which can be taken as:

$$K_{mat}^2 = 1000\frac{E}{A}(C_v - 17.6)$$

where:

k_{mat} = material toughness, in N/mm^{3}/^{2}

C_v = Charpy energy, in J

E = Young's modulus, in N/mm^2

A = section area for Charpy test, in mm^2, normally $A = 80$ mm^2

The geometry function Y can be expressed as:

$$Y = \frac{F}{\sqrt{Q}} \cdot \left[1 - 1.8 \cdot \left(\frac{D_d}{D}\right) + 5.1 \cdot H \cdot \left(\frac{D_d}{t}\right)\right]$$

where:

D_d = depth of dent

and geometry correction factors Q, F and H are given by the following:

$$Q = 1 + 1.464\left(\frac{a}{c}\right)^{1.65} \quad for \; \frac{a}{c} \le 1$$

where:

c = half length of dent

$F = \left[M_1 + M_2\left(\frac{a}{t}\right)^2 + M_3\left(\frac{a}{t}\right)^4\right]f_\phi g f_w$

where:

$M_1 = 1.13 - 0.09\frac{a}{c}$

$M_2 = -0.54 + \frac{0.89}{0.2+(\frac{a}{c})}$

$M_3 = 0.5 - \frac{1}{0.65+(\frac{a}{c})} + 14\left(1 - \frac{a}{c}\right)^{24}$

$g = 1 + \left[0.1 + 0.35\left(\frac{a}{t}\right)^2\right](1 - \sin\phi)^2$

where:

ϕ = parametric angle of the elliptical crack

The function f_ϕ, an angular function from the embedded elliptical-crack solution is:

$$f_\phi = \left[\left(\frac{a}{c}\right)^2 \cos^2\phi + \sin^2\phi\right]^{\frac{1}{4}}$$

The function f_w, a finite-width correction factor is:

$$f_w = \left[\sec\left(\frac{c}{D} \cdot \sqrt{\frac{a}{t}}\right)\right]^{\frac{1}{2}}$$

The function H has the form:

$$H = H_1 + (H_2 - H_2)\sin^P\phi$$

where:

$$p = 0.2 + \left(\frac{a}{c}\right) + 0.6\left(\frac{a}{t}\right)$$

$$H_1 = 1 - 0.34\left(\frac{a}{t}\right) - 0.11\left(\frac{a}{c}\right)\left(\frac{a}{t}\right)$$

$$H_2 = 1 + G_1\left(\frac{a}{t}\right) + G_2\left(\frac{a}{t}\right)^2$$

$$k_{fab} =$$

and

$$G_1 = -1.22 - 0.12\left(\frac{a}{c}\right)$$

$$G_2 = 0.55 - 1.05\left(\frac{a}{c}\right)^{0.75} + 0.47\left(\frac{a}{c}\right)^{1.5}$$

Design Recommendations for Subsea LNG Pipelines

16

16.1 General

The design, fabrication, installation, testing and maintenance of subsea liquefied natural gas (LNG) pipelines are to be based on all applicable requirements of Chaps. 1 through 4. Alternatively, industry acceptable codes and standards for subsea pipeline systems may also be used based upon mutual agreement between Class and the asset designer/owner. However, the criteria for pipelines operating under ambient temperatures may be inadequate for pipelines working under cryogenic temperatures. This chapter provides special recommendations for subsea LNG pipelines working under cryogenic conditions.

16.1.1 Application

This chapter sets out recommendations for technical aspects specific to metallic subsea cryogenic pipelines transporting LNG from offshore LNG receiving terminals to LNG storage facilities or vice versa, or subsea LNG pipelines installed in lieu of the conventional trestle/jetty construction. This chapter is not applicable to in-air offshore LNG transfer pipes or hoses.

© The Author(s), under exclusive license to Springer Nature Switzerland AG 2025
A. A. Olsen, *Subsea Pipeline Systems*, Synthesis Lectures on Ocean Systems
Engineering, https://doi.org/10.1007/978-3-031-74790-8_16

16.2   Materials and Welding

16.2.1 Materials

Materials suitable for cryogenic services are to have sufficient toughness and strength at the lowest anticipated temperature during operation or testing. Metals that can be used for cryogenic services include, but are not limited to, austenitic stainless steels 304L and 316L, Invar (with 36% content of nickel alloy), and aluminium alloy 5083. Components (such as outer pipe in a pipe-in-pipe design) which remain at ambient temperatures during operation are allowed to be manufactured from carbon steel. The selection of materials is also to consider thermal expansion properties and possible temperature-induced strain (thermal strain) in pipe joints and connections. Considerations are also to be given to the requirements specified in Chap. 2 for the selection of materials, welding, components, and corrosion control. Materials for an insulation system are to be designed with sufficiently low thermal conductivity, watertight and enough mechanical strength for handling.

16.2.2 Welding

The welding of metallic LNG pipes is to be performed in accordance with welding procedures specified in API STD 1104, Section IX of the ASME Boiler and Pressure Vessel Code or comparable industry-acceptable standards. The welds are to be inspected visually, and by non-destructive testing procedures that are developed and qualified according to API STD 1104.

16.3   Design Considerations

16.3.1 General

The identification, definition, and determination of loads to be considered in the design of subsea LNG pipelines are to be in accordance with Chap. 5. Chapter 8 is applicable for strength and stability criteria including hoop stress, longitudinal stress, Von Mises stress, global and local buckling, and fatigue. Special design considerations for subsea LNG pipelines are listed in the following paragraphs.

16.3.2 Loads

The subsea LNG pipeline system is to be designed to withstand the following loads:

- Loads induced by the installation process (string weight, associated tension, hydrostatic pressure, and temperature variations),
- Cryogenic bending during fill-up/emptying of the line (thermal loads due to temperature differential across cross-sections during fill-up/emptying), and
- Loads in operation, including thermal loads, internal and external pressure load, seabed interference, and fatigue load during the life of the pipeline.

16.3.3 Allowable Stresses

Criteria for allowable stresses are to be in accordance with Chap. 5. Consideration is to be given to changes in material tensile properties from ambient temperatures to cryogenic temperatures. Generally, metallic materials have better tensile properties at lower temperatures. However, increase of allowable stresses may be considered only if the safety factors specified in Chap. 5 can be strictly maintained for all load cases and is subject to approval from Class.

16.3.4 On Bottom Stability

Subsea LNG pipelines resting on the seabed, trenched, or buried are not to move from their as-installed position unless accounted for in the design. The lateral stability of the pipelines may be assessed in the design using two-dimensional static or three-dimensional dynamic analysis methods specified in Chap. 9.

16.3.5 Cryogenic Bending

Fill-up and emptying procedures of subsea LNG pipelines are to be clearly defined. The fill-up or emptying processes could be critical in terms of static and fatigue strengths. Differential cooling or heating across a cross-section of the pipe may result in cryogenic bending (significant pipe deformation or upheaval movement). Heat transfer analysis is to be performed to evaluate the effect of cooling rate on pipe stresses at filling.

16.3.6 Global and Local Buckling

Subsea LNG pipelines designed and fabricated without internal bellows or expansion loops over long distances may be subject to global buckling due to internal overpressure and differential temperature across the cross-section of the pipe during operation. Analysis is to be performed to predict the possible position and amplification of the buckle. Local buckling may occur when the pipeline is under external overpressure, bending or their combination. The formulas in Chap. 8 are to be used for the prediction of local buckling. The maximum anticipated differential pressure and bending moment acting on the pipe are to be used in the formulas. In the case of pipe-in-pipe designs, accidental flooding of the annulus is to be accounted for local buckling of the internal pipe in a damaged condition.

16.3.7 Fatigue

Criteria for fatigue assessment of subsea LNG pipelines are specified in Chap. 8. Subsea pipelines for cryogenic services may be subject to high thermal stresses throughout their entire life cycle. Considerations should be given to the temperature-induced high-stress range, low-cycle fatigue. Temperature effects on material fatigue resistance properties are to be assessed in choosing an S–N design curve. VIV induced fatigue damages are to be accounted for pipelines subjected to VIV effects. A safety factor of 10 is to be used in calculating fatigue life of subsea LNG pipelines.

16.3.8 Insulation

Subsea LNG pipelines are to be insulated such that the outermost pipe maintains a positive surface temperature at all times. No build-up of ice is allowed on any components of the pipelines exposed to seawater, except if it can be proved that ice build-up has no adverse effects on integrity of the pipeline. Effective thermal conductivity (U-value) of the pipeline is to be designed such that two-phase flow is avoided in the LNG pipelines. In addition, consideration should be given to the amount of heat gain between ship unloading operations. The boil-off rate of LNG inside the pipeline is to be controlled within an allowable level defined by the asset owner/designer.

16.3.9 Thermal and Mechanical Analysis

Finite element analysis using an industry-accepted solver is recommended for thermal calculation and mechanical analysis of the pipeline components. The mechanical analysis is to consider temperature distributions from thermal calculations and structural and contained fluid weight.

16.4   Installation

Subsea LNG pipelines can be installed using installation methods described in Chap. 10. In the case of installation by towing, the pipeline system can be launched into the sea from either a perpendicular site or a parallel site depending on the type of fabrication site available. Stresses that may be induced in the pipe during lay operations are to be fully considered. For pipes that will remain bent, initial as-laid stresses are to be accounted for in any in-place analysis. Once the towing operation has been completed and the pipeline has been installed into its final position on the seabed, the pipeline is to be secured by trenching, backfilling, protective covering or anchoring.

16.5   Testing

16.5.1 Mill Pressure and Thermal Test

The mill pressure test is required to constitute a pressure containment proof test and ensure that all pipe sections have at least a minimum yield stress. Thermal cycling tests and boil-off are recommended for LNG pipelines involving novel design concepts.

16.5.2 Test After Construction

Pressure testing is to be performed at the ambient temperature on the completed LNG pipeline system and on all components not tested comprising the pipeline system, or components requiring a higher test pressure than the remainder of the pipeline. Testing procedure, test media, test pressure level and acceptance criteria are to be in accordance with Chaps. 10 and 12.

Correction to: Subsea Pipeline Systems

Correction to:

A. A. Olsen, *Subsea Pipeline Systems*, Synthesis Lectures on Ocean Systems Engineering,
https://doi.org/10.1007/978-3-031-74790-8

This book contains overlap in text with the previously published content [1] that was inadvertently omitted. The authors failed to attribute the reference [1]. The authors have now obtained permission to re-use this content from the American Bureau of Shipping.

Where [1] is: American Bureau of Shipping (2024), Rules and Guides https://ww2.eagle.org/en/rules-and-resources/rules-and-guides.html

---

The updated version of this book can be found at
https://doi.org/10.1007/978-3-031-74790-8

Glossary

Accommodation Areas

Accommodation Areas as used in this book refers to and includes the "accommodation spaces", the "public areas" and the "service spaces" that are within or directly adjacent to the "accommodation spaces" or "public spaces".

Accommodation Spaces

Accommodation Spaces as used in this book are those spaces used for corridors, lavatories, cabins, offices, hospitals, cinemas, game and hobby rooms, barber shops, pantries containing no cooking appliances and similar spaces.

Chemical Carrier

A ship constructed or adapted for the carriage in bulk of any liquid product listed in Chapter 17 of the IMO International Bulk Chemical Code (IBC Code).

Classified Area

A location in which flammable gases or vapours are or may be present in the air in quantities sufficient to produce explosive or ignitable mixtures (refer to API RP 500, or API RP 505 for additional details). See also "Hazardous Area".

Container Carrier

A vessel designed primarily for the carriage of containers in holds or on deck or both, with structures for that purpose, such as cell guides, pedestals, etc.

Dry Bulk Cargo Carrier

A ship that is constructed generally with single deck, topside tanks and hopper side tanks in cargo spaces, and is intended primarily to carry dry cargo in bulk. It includes a vessel of such type as ore carrier or combination carrier.

A. A. Olsen, *Subsea Pipeline Systems*, Synthesis Lectures on Ocean Systems
Engineering, https://doi.org/10.1007/978-3-031-74790-8

Escape Route
This is a designated path used by personnel to evade an immediate danger and ultimately leads to a temporary refuge or muster station.

Fixed Installation
A bottom-fixed offshore facility permanently affixed to the sea floor. The term includes, but is not limited to, fixed platforms, guyed towers, jack-ups elevated at same location longer than five (5) years, converted fixed installations, etc.

Flammable Fluid
Any fluid, regardless of its flash point, capable of feeding a fire is to be treated as a flammable fluid. Aviation fuel, diesel fuel, hydraulic oil (oil based), lubricating oil, crude oil and hydrocarbons are to be considered flammable fluids.

Flash Point
The minimum temperature at which a combustible liquid gives off vapor in sufficient concentration to form an ignitable mixture with air near the surface of the liquid or within the vessel used, as determined by the test procedure and apparatus specified in NFPA 30. Ignitable mixture means a mixture that is within the flammable range (between the upper and lower limits) and is therefore capable of propagation of flame away from the source of ignition.

Floating Installation
An offshore facility designed to provide hydrocarbon processing and/or hydrocarbon storage and offloading of hydrocarbons. The term Floating Installation is used to generically identify a buoyant facility that is site-specific. This installation is securely and substantially moored so that it cannot be moved without a special effort. The term includes, but is not limited to, Tension Leg Platforms (TLP), Spar Buoys, Permanently Moored Shipshape Hulls and Semisubmersibles.

General Cargo Carrier
A vessel constructed for the transportation of suitable cargoes in compatible combinations and packaging, such as boxes, bales, barrels, drums, etc.

Hazardous Area
A location in which flammable gases or vapours are or may be present in the air in quantities sufficient to produce explosive or ignitable mixtures (refer to API RP 500, or API RP 505 for additional details). See also "Classified Area".

Industrial Area

An area that is intended to primarily serve the industrial function and operation of a MODU or offshore installation. This includes the drilling area for MODUs and the areas containing the process and process support systems on offshore installations.

Jet Fire

A fire resulting from the combustion of a flammable liquid or gas that is being continuously released with some significant momentum in a particular direction or directions.

Liquefied Gas Carrier

A vessel designed and constructed for the transportation in bulk of liquefied gas, or other products listed in Chapter 19 of the IMO International Code for the Construction and Equipment of Ships Carrying Liquefied Gases in Bulk (International Gas Carrier Code), or Chapter XIX of the IMO Code for the Construction and Equipment of Ships Carrying Liquefied Gases in Bulk (Gas Carrier Code).

Local Fixed Fire Extinguishing System

The fixed fire extinguishing system installed to protect a particular piece of equipment.

Lower Explosive Limit (LEL)

The lowest concentration of combustible vapours or gases, by volume in mixture with air, which can be ignited at ambient conditions.

Machinery Spaces

Machinery Spaces are all "Machinery Spaces of Category A" as well as all other machinery spaces containing propulsion machinery, boilers, oil fuel units, steam and internal combustion engines, generators and major electrical machinery, oil filling stations, refrigerating, stabilising, ventilation and air conditioning machinery, and similar spaces and trunks to such spaces.

Machinery Spaces of Category A

Machinery Spaces of Category A are those spaces and trunks to such spaces which contain either:

(1) Internal combustion machinery used for main propulsion;
(2) Internal combustion machinery used for purposes other than main propulsion where such machinery has an aggregate total power output of not less than 375 kW (500 hp);
(3) Any oil-fired boiler, oil fuel unit, or any oil-fired equipment other than boiler, such as inert gas generator, incinerator, waste disposal unit, etc.

Main Fixed Fire Extinguishing System

The fixed fire extinguishing system installed to protect the entire machinery space.

Mobile Offshore Drilling Unit (MODU)

A mobile offshore unit of the self-elevating or column-stabilised type, that is fitted with drilling equipment and utilised to carry out drilling operations.

Mobile Offshore Unit (MOU)

A mobile offshore unit of self-elevating or column-stabilised type, not fitted with drilling equipment, production facilities, hydrocarbon storage, or any other system onboard handling hydrocarbons.

Mud Gas Separator Unit

A Mud Gas Separator Unit is intended to separate large volumes of free gas from within the drilling fluid. Also referred to as a gas-buster.

Oil Carrier

A ship constructed or adapted primarily to carry oil and/or oil product in bulk and includes combination carriers.

Oil Fuel Unit

An Oil Fuel Unit is defined as any equipment, such as pumps, filters and heaters, used for the preparation and delivery of fuel oil to oil-fired boilers (including incinerators and inert gas generators), internal combustion engines or gas turbines at a pressure of more than 1.8 kgf/cm^2 (0.18 N/mm^2, 26 psi).

Passenger

Every person other than:

1. The master and the members of the crew or other persons employed or engaged in any capacity on board a ship on the business of that ship; and
2. A child under one year of age.

Oil Systems, Systems Containing Oil

As used in this Chapter, the terms Oil Systems or Systems Containing Oil refer to any piping system or equipment that contains or transfers flammable fluids.

Pool Fire

A fire resulting from the combustion of the vapours above a horizontal body of a flammable fluid that has zero or low initial momentum and is typically contained by some surrounding structure.

Process Area
Area where processing equipment is located. This includes wellhead/manifold areas.

Process Support Systems
Utility and auxiliary systems that complement the hydrocarbon production and process systems. These systems do not directly handle hydrocarbons.

Public Spaces
Portions of the accommodations which are used for halls, dining rooms, lounges, and similar permanently enclosed spaces.

Ro-Ro Cargo Spaces
Spaces not normally subdivided in any way and extending to either a substantial length or the entire length of the vessel in which goods [packaged or in bulk, in or on road cars, vehicles (including road tankers), trailer containers, pallets, demountable tanks or in or on similar stowage units or other receptacles] can be loaded and unloaded normally in a horizontal direction.

Service Spaces
Spaces used for galleys, pantries containing cooking appliances, lockers, mail and specie rooms, storerooms, workshops other than those forming part of the machinery spaces, and similar spaces and trunks to such spaces.

Special Personnel
All persons who are not passengers or members of the crew or children of under one year of age and who are carried on board in connection with the special purpose of that vessel or because of special work being carried out aboard that vessel.

Special Purpose Vessel
A vessel which by reason of its function carries on board more than twelve (12) special personnel.

Vehicle Carrier
A vessel designed and constructed for the carriage of vehicles only.

Appendix

Strength and Stability Criteria

General

References by Organization

Standards/codes acceptable to the Bureau are not limited to the following references.

When updates of the referenced documents are available, these are to be used as far as possible.

Which standards/codes to be followed during design, manufacturing, transportation, storage, installation, system testing, operation, amendment, decommission etc. isgenerally to follow international agreements and be agreed upon between Local Authorities, Owners, Operators, Clients and Contractors.

The Bureau claims the right to reject documents, procedures etc. where standards/codes are judged misused, e.g., by "shopping around".

ABS

American Bureau of Shipping
Code No. Year/Edition Title
Latest Edition Rules for Building and Classing Single Point Moorings
Latest Edition Rules for Building and Classing Offshore Installations
Latest Edition Rules for Building and Classing Floating Production Installations
Latest Edition Rules for Building and Classing Marine Vessels
Latest Edition Guide for the Fatigue Assessment of Offshore Structures
Latest Edition Guide for Building and Classing Subsea Pipeline Systems

AGA

American Gas Association
Code No. Year/Edition Title
PR-3-805 1989 A Modified Criterion for Evaluating the Remaining Strength of Corroded Pipe

© The Editor(s) (if applicable) and The Author(s), under exclusive license
to Springer Nature Switzerland AG 2025
A. A. Olsen, *Subsea Pipeline Systems*, Synthesis Lectures on Ocean Systems
Engineering, https://doi.org/10.1007/978-3-031-74790-8

L51698 1993 Submarine Pipeline On-Bottom, Vol. I:Stability Analysis and Design Guidelines

AISC

American Institute of Steel Construction
Code No. Year/Edition Title
1989 ASD Manual of Steel Construction, 9th Edition

API

American Petroleum Institute
Code No. Year/Edition Title
SPEC 5L 2000 Specification for Line Pipe
SPEC 5LC 1998 CRA Line Pipe
SPEC 5LD 1998 CRA Clad or Lined Steel Pipe
SPEC 5LW 1996 Recommended Practice for Transportation of Linepipe on Barges and Marine Vessels
SPEC 6A 1999 Wellhead and Christmas Tree Equipment
SPEC 6D 1996 Pipeline Valves (Gate, Plug, Ball, and Check Valves)
SPEC 17D 1992 Subsea Wellhead and Christmas Tree Equipment
STD 600 1997 Steel Gate Valves-Flanged and Butt-Welding Ends, Bolted and Pressure Seal Bonnets
STD 1104 1999 Welding of Pipelines and Related Facilities,
RP 2A-WSD 2000 Planning, Designing, and Constructing Fixed Offshore Platforms – Working Stress Design
RP 2N 1995 Planning, Designing, and Constructing Structures and Pipelines for Arctic Conditions
RP 2RD 1998 Design of Risers for Floating Production Systems (FPS's) and Tension-Leg Platforms (TLP's)
RP 2SK 1996 Design and Analysis of Stationkeeping Systems for Floating Structures
RP 2T 1997 Planning, Designing and Constructing Tension Leg Platforms
RP 5L1 1996 Railroad Transportation of Line Pipe
RP 5LW 1996 Recommended Practice for Transportation of Line Pipe on Barges and Marine Vessels
RP 14G 2000 Fire Prevention and Control on Open Type Offshore Production Platforms.
RP 17A 1996 Design and Operation of Subsea Production Systems
RP 579 2000 Fitness-For-Service
RP 1110 1997 Pressure Testing of Liquid Petroleum Pipelines
RP 1111 1999 Design, Construction, Operation, and Maintenance of Offshore Hydrocarbon Pipelines

ASME

American Society of Mechanical Engineers
Code No. Year/Edition Title
B16.5 1996 Pipe Flanges and Flanged Fittings
B16.9 1993 Factory-Made Wrought Steel Buttwelding Fittings
B16.10 2000 Face to Face and End to End Dimensions of Valves
B16.11 1996 Forged Steel Fittings, Socket-Welding and Threaded
B16.20 1998 Metallic Gaskets for Pipe Flanges: Ring-Joint, Spiral-Wound, and Jacketed
B16.25 1997 Buttwelding Ends
B16.34 1996 Valves-Flanged, Threaded, and Welding End
B31G 1991 Manual for Determining the Remaining Strength of Corroded Pipelines: A supplement to B31
B31.4 2002 Pipeline Transportation Systems for Liquid Hydrocarbons and Other Liquids
B31.8 1999 Gas Transmission and Distribution Piping Systems
Boiler and Pressure Vessel Code 1998 Section V: Non-Destructive Examination
Section VIII: Pressure Vessels –Divisions 1 and 2 Section IX: Welding and Brazing Qualifications

ASNT

American Society for Nondestructive Testing
Code No. Year/Edition Title
A3.0 2001 Standard Welding Terms and Definitions

ASTM

American Society for Testing and Materials
Code No. Year/Edition Title
A36 2003 Specification for Structural Steel
A82 2002 Specification for Steel Wire, Plain, for Concrete Reinforcement
A105 2003 Specifications for Forgings, Carbon Steel for Piping Components
A193M 2003 Specification for Alloy-Steel and Stainless Steel Bolting Materials for High Temperature Service
A194M 2003 Specification for Carbon and Alloy-Steel Nuts for Bolts for High- Pressure and High-Temperature Service
A216 2003 Specification for Steel Castings, Carbon suitable for Fusion Welding for High Temperature Service
A234 2003 Specification for Piping Fittings of Wrought Carbon Steel and Alloy Steel for Moderate and Elevated Temperatures
Code No. Year/Edition Title
A370 2003 Test Methods and Definitions for Mechanical Testing of Steel Products
A388 2003 Practice for Ultrasonic Examination of Heavy Steel Forgings

A578 2001 Specification for Straight-Beam Ultrasonic Examination of Plain and Clad Steel Plates for Special Application

A694 2003 Specification for Forgings, Carbon and Alloy Steel for Pipe Flanges, Fittings,Valves, and Parts for High Pressure Transmission Service

A790/ A790M-03 2003 Standard Specification for Seamless and Welded Ferritic/ Austenitic Stainless Steel Pipe

B861 2003 Standard Specification for Titanium and Titanium Alloy Seamless Pipe

B862 2002 Standard Specification for Titanium and Titanium Alloy Welded Pipe

C29 2003 Test Method for Unit Weight and Voids in Aggregate

C31 2003 Test Method for Making and Curing Concrete Test Specimens in the Field

C33 2003 Specifications for Concrete Aggregates

C39 2003 Test Method for Compressive Strength of Cylindrical Concrete Specimens

C150 2002 Specifications for Portland Cement

C172 1999 Method of Sampling of Fresh Mixed Concrete

C642 1997 Test Method for Specific Gravity, Absorption and Voids in Hardened Concrete

D75 2003 Methods for Sampling Aggregates

D4285 1999 Method for Indicating Oil or Water in Compressed Air

E23 2002 Methods for Notched Bar Impact Testing of Metallic Materials

E165 2002 Practice for Liquid Penetrant Inspection Method

E337 2002 Test Method for Measuring Relative Humidity with a Psychrometer (The Measurement of Wet and Dry Bulb Temperatures)

AWS

American Welding Society
Code No. Year/Edition Title
A3.0 2001 Standard Welding Terms and Definitions
A5.5 1996 Specification for Low Alloy Steel Electrodes for Shielded Metal Arc Welding
D1.1 2000 Structural Welding Code-Steel
D3.6M 1999 Specification for underwater welding
American Water Works Association (AWWA)
Code No. Year/Edition Title
AWWA C-203 2002 Coal-Tar Protective Coatings and Linings for Steel Water Pipelines, Enamel and Tape Hot-Applied (Modified for use of Koppers Bitumastic High-Melt Enamel)

BSI

British Standards Institute
Code No. Year/Edition Title
BS 427 1990 Method for Vickers hardness test and for verification of Vickers hardness testing machines

BS 4147 1980 Bitumen Based Hot Applied Coating Materials for Protecting Iron and Steel

BS 4515-1 2000 Specification for welding of steel pipelines on land and offshore. Carbon and carbon manganese steel pipelines

BS 4515-2 1999 Specification for welding of steel pipelines on land and offshore. Duplex stainless steel pipelines

BS 7910 1999 Guidance on Methods for Assessing Acceptability of Flaws in Fusion Welded Structures

BS 8010-3 1993 Code of practice for pipelines. Pipelines subsea: design, construction and installation

CSA

Canadian Standard Association
Code No. Year/Edition Title
Z662-99 1999 Oil and Gas Pipeline Systems

HSE

Health and Safety Executive
Code No. Year/Edition Title
1990 Offshore Installations: Guidance on Design, Construction and Certification

ISO

International Organization of Standards
Code No. Year/Edition Title
ISO 3183-1 1996 Petroleum and natural gas industries- Steel pipe for pipelines-Technical delivery conditions, Part 1: Pipes of Requirement Class A
ISO 3183-2 1996 Petroleum and natural gas industries- Steel pipe for pipelines-Technical delivery conditions, Part 2: Pipes of Requirement Class B
Code No. Year/Edition Title
ISO 3183-3 1999 Petroleum and natural gas industries- Steel pipe for pipelines-Technical delivery conditions, Part 3: Pipes of Requirement Class C
ISO 9712 1999 Nondestructive testing Qualification and certification of personnel

MSS

Manufacturers Standardization Society of the Valve and Fittings Industry, Inc
Code No. Year/Edition Title
SP-6 2001 Standard Finishes for Contact Faces of Pipe Flanges and Connecting- End Flanges of Valves and Fittings
SP-44 1996 Steel Pipe Line Flanges
SP-53 2002 Quality Standard for Steel Castings and Forgings - Magnetic Particle Examination Method.

SP-54 2002 Quality Standard for Steel Castings—Radiographic Examination Method
SP-55 2001 Quality Standard for Steel Castings—Visual Method
SP-75 1998 Specification for High Test Wrought Butt Welding Fittings

NACE International
Code No. Year/Edition Title
MR0175 2000 Sulfide Stress Cracking Resistant Metallic Materials for Oil Field Equipment
RP0176 1994 Corrosion Control of Steel, Fixed Offshore Platforms Associated with Petroleum Production
RP0274 1998 Recommended Practice, High Voltage Electrical Inspection of Pipeline Coatings Prior to Installation
RP0387 1999 Metallurgical and Inspection Requirements for Cast Sacrificial Anodes for Offshore Applications.
RP0675 Withdrawn, new standard planned. Recommended Practice for Control of Corrosion on Offshore Steel Pipelines.
TM0177 1996 Laboratory Testing of Metals for Resistance to Sulfide Stress Cracking and Stress Corrosion Cracking in H2S Environments
TM0284 1996 Evaluation of Pipeline and Pressure Vessel Steels for Resistance to Hydrogen Induced Cracking

SSPC

The Society for Protective Coatings
Code No. Year/Edition Title
SSPC-PA-2 1996 Measurement of Dry Paint Thickness With Magnetic Gauges
SSPC-SP-1 2000 Solvent Cleaning
SSPC-SP-3 2000 Power Tool Cleaning
SSPC-SP-5 2000 White Metal Blast Cleaning
SSPC-SP-10 2000 Near-white Blast Cleaning

SIS

Swedish Standard Institute
Code No. Year/Edition Title
SIS 05 5900 1988 Pictorial Surface Preparation Standard for Painting Steel Surfaces